OUR FRAGILE
WATER PLANET

OUR FRAGILE
WATER PLANET

An Introduction to the Earth Sciences

C. L. Mantell

Professor Emeritus
New Jersey Institute of Technology
and
Consulting Chemical Engineer

and

A. M. Mantell

PLENUM PRESS · NEW YORK AND LONDON

Library of Congress Cataloging in Publication Data

Mantell, Charles Letnam, 1897-
    Our fragile water planet.

    Includes bibliographical references and index.
    1. Earth Sciences. 2. Environmental and protection. I. Mantell, Adelaide M.,
1899-   joint author. II. Title.
QE33.M316   550   76-20730
ISBN 0-306-30877-0

PREFACE

This volume is an introduction to the study of the earth sciences, a multitude which embraces geochemistry, the science of the earth's composition; geology, the science of the earth's structure; meteorology and climatology, the study of both local and planetary weather; tectonics, the fledgling science of the movement of sections of the earth, of earthquakes, and of volcanoes; biology and agricultural engineering; the water cycle and reclamation; the chemistry of the atmosphere and the origin of the changes it undergoes; the seas, the oceans, or oceanography; beach movement and deserts; hydrology, the science of water from the viewpoint of the sources of energy; pressure and temperature effects; the crust or lithosphere; the hydrosphere, or water areas; the atmosphere, winds, weather, hurricanes, and cyclones, as well as the technology of tracking them; the interfaces of the sections of the planet, and the uses of the earth by its peoples.

The human race has developed on the crust and at the interfaces of the land, air, and sea of an unstable planet — a dynamic geological entity whose thermal equilibrium is still millions of years away. The crust of the earth in its movement and cracking evolves earthquakes and volcanoes which are destructive of human work and limit our habitation. Over the last 1500 years, earthquakes have killed as many people as now inhabit the planet, and fifteen times the present U.S. population. The sun is the fundamental source of our planet's energy and will remain so as far into the future as we can see. Both these internal and external energies, which so abound, are coming increasingly under human control, as they must to assure our survival.

Our species is an oxygen-breathing animal, exhaling carbon dioxide, in contrast to plants, trees, algae, and the like, which inhale carbon dioxide and, using sunlight for energy, return oxygen to the atmosphere. Human existence is dependent on the water cycle, the evaporation of water from the seas to vapor-forming clouds, which in turn fall as rain, snow, hail, or sleet. We have progressively learned to correct unevenness in the natural distribution of water to better manage our crops and animals.

We have learned to manage our farms, food production, forests, and to a degree the atmosphere, the land, and the seas. When we are unable or find it uneconomic to use some lands, we often satisfy our esthetic sensibilities by creating from them national, state, and regional parks. We are enabled by our great technology to reach every corner of earth by sea, air, or land, and to enjoy instant communication with our fellow beings over all the world.

We still seek shelter from hurricanes and cyclones; we fight forest fires; we protect ourselves against floods, sometimes insufficiently; and we bend before the force of the sea which moves our beaches, promentories, and bays.

Our greatest enemy is the insect world and its hordes, which wage war on our crops, our domesticated animals, our structures, our forests, and our very bodies by being carriers of disease.

In the 20th century, our science has divided bacteria, fungi, and other microorganisms into pathogens, against which we make deadly crusade, and nonpathogens, which we use for controlled fermentation, biologic oxidation of wastes, and food preparation. We have learned to prepare antibodies, toxins, and antitoxins in the 100 years since discovering the germ theory of disease.

In our agricultural engineering, we have learned of genes, genetics, the development of better plants, trees, hybrid wheat and corn, so that now one farmer feeds nearly 50 people. Through these means we have so humanized the world in which we live that it can be accurately said we manage it.

Further evidence of our humanizing efforts lies very clearly in our growing conservationism in the mountains, forests, and lakes, in the innumerable national monuments, forests, and state parks treasured in every state in the Union, and elsewhere in such foresighted projects as the animal parks found extensively in Africa.

We have taken minerals, metals, and fossil fuels (such as natural gas, coal, and petroleum) from nature's burial grounds and learned how to use these energy forces to extend our managed domain. We have crossed rivers with our bridges, or tunnelled under them, and provided highways wherever we wanted to go. We have travelled to, and walked on, the moon.

We have better health than our ancestors, and our life cycle increases with each generation. We continue to grow in the technical, engineering, and scientific knowledge necessary to exercise beneficient dominion over our fragile water planet.

C. L. Mantell

Munsey Park
Manhasset, New York

Contents

1 Introduction 1

  1-1   The Earth and the Solar System 1
  1-2   The Energy Source: The Sun 2
  1-3   Pressure and Temperature Effects 4

2 The Crust (Lithosphere) 7

  2-1   The Structure of the Earth 7
  2-2   Continental Drift and Tectonics 7
  2-3   The Andes and Mountain Building 13
  2-4   Earthquakes 19
  2-5   Volcanoes 28
  2-6   Composition of the Crust 36
  2-7   The Rocks We Use 38
  2-8   Karst Regions 39

3 The Hydrosphere 41

  3-1   The Water Cycle 41
  3-2   The Great Lakes of the United States 48
  3-3   Rivers and Flood Control 53
  3-4   The Tennessee Valley Authority and Floods 57
  3-5   Drainage of Wetlands 63
  3-6   Glaciers 69
  3-7   Deserts 71
  3-8   Reclamation 76
  3-9   Geysers 80

4 The Atmosphere 87

  4-1   Composition 87
  4-2   Weather 90

4-3   Hurricanes   90
4-4   Cyclones   92
4-5   Tornadoes..   94
4-6   Winds   97

5   The Interfaces 103

5-1   Islands.. 103
5-2   Beaches 103
5-3   El Niño 112
5-4   Swamps and Marshes... 116
5-5   Reefs 118
5-6   Red Tides 118
5-7   Subsidences 119

6   Chemical Cycles 123

6-1   The Carbon Cycle and the Biosphere 123
      6-1a   Photosynthesis: Carbon Dioxide and Oxygen 123
      6-1b   Carbon Monoxide 127
      6-1c   Methane and Natural Gas 130
      6-1d   Cellulose and Coal 131
      6-1e   Petroleum... 132
6-2   The Inorganic Element Cycles 138
      6-2a   Nitrogen.. 138
      6-2b   Sulfur and Sulfur Dioxide 138
      6-2c   Phosphorus 142
      6-2d   Calcium 143
      6-2e   Metals 143
      6-2f   Chlorine.. 144
      6-2g   Silicon 145

7   Humanizing the Earth 147

7-1   Agricultural Engineering... 147
7-2   Plagues and Insecticides 153
7-3   Forests and Insects.. 157
7-4   Fires 167

8   Conservation 171

8-1   The National Parks and Monuments 171
8-2   African Game Lands 181
8-3   The Lakes and Parks of California 184

8-4 The Everglades 194
8-5 Avery Island, Louisiana 197

9 Civilization in the Twentieth Century 199

9-1 Our Waste in an Urban Environment 199
9-2 Technological Growth 209

Index 219

INTRODUCTION

1–1 THE EARTH AND THE SOLAR SYSTEM

Our fragile water planet is a unit within a solar system, revolving around the sun. The solar system embraces the sun, the planets and their satellites (moons), the asteroids, the comets, and meteorites. The sun is only one star in our galaxy. The galaxy has the shape of a lens, embraces more than 10^{11} (or 100,000,000) stars,[1] and has a diameter of about 70,000 light-years.

Light travels at 186,000 (or 186×10^3) miles per second; with 60 seconds to the minute, 60 minutes to the hour, that is, 3600 seconds to the hour, the speed of light is $(186 \times 10^3) \times (3.6 \times 10^3)$ or 669.6×10^6 (that is, 669,600,000) miles per hour. A light-year is the distance traversed in 8760 ($= 24 \times 365$) hours. This equals approximately 6×10^{12} (that is, 6,000,000,000,000 or six quadrillion) miles, or, in metric terms, 10^{13} km.

Beyond our galaxy there are a very large number of other systems of approximately the same size. The nearest one is the Andromeda nebula, about 1 million light-years away.

If an astronaut were to travel at the supersonic speed of 1000 miles per hour to a destination one light-year from the earth, a week's travel (168 hours) would carry him 168×10^3 miles. In one year (8760 hours) 8.7×10^6 miles would be covered. By the time one light-year had been traversed, the traveler would be 669,600 years old. If family life were possible on this journey, and counting four generations per 100 years, the traveler completing the journey would be over 25,000 generations removed from the original voyager.

Travel at a rate of 1000 miles per second, the one-hundred-and-eighty-sixth part of the speed of light, would permit traversal of one light-year in 186

---

[1]The powers of 10 concept is useful in dealing with large numbers: $10^1 = 10$; $10 \times 10 = 10^2 = 100$; $10 \times 10 \times 10 = 10^3 = 1000$; $1000 \times 1000 = 10^3 \times 10^3 = 10^6 = 1,000,000$; $10^3 \times 10^3 \times 10^3 = 10^9 = 1,000,000,000$; etc.

years. The surviving traveler would be the seventh or eighth descendant of the one who started the voyage.

Geophysicists and geochemists estimate the age of the universe to be 2×10^9 years, equivalent to 8×10^7 generations of the human race.

Geologists, geophysicists, astronomers, and seismologists appear to agree that the planet has an inner core, a mantle surrounding this, an outer crust, and a gaseous atmosphere. The crust is not uniform but heterogeneous, as the diverse forms of the oceans, lakes, valleys, mountains, and the deep ocean basins demonstrate. Obviously, the crust varies in thickness from place to place. Over the whole earth there is a discontinuous skin of sedimentary and metamorphic rocks, which may be 50 km thick, or more, in geosynclinal belts, but which is insignificant when compared with the mass of the crust.

There appears to be a discontinuity (called the Mohorovičić discontinuity) separating the heterogeneous crust from the more homogeneous mantle. At this discontinuity, the velocity of seismic (explosive) waves suddenly increases. In 1873, J. D. Dana, a mineralogist, from a study of meteorites, suggested that our planet had an iron core. There is a second seismic discontinuity between the mantle and the core at 2900 km.

Table 1 gives the composition, the mass or weight, the thickness, and the volume of the atmospheric or gaseous envelope; the biosphere where we live and have adapted ourselves, and where all other life, in the main, propagates; the hydrosphere of the oceans, lakes, rivers, glaciers, and snow belts (the habitat of marine life); the rocky crust; the mantle of homogeneous silicate rock; and the siderosphere or center of the solid core, surrounded by molten metal or metal compounds.

The area congenial to life is extremely small, and the interface of land, sea, and air is not a static system, but rather a dynamic one that undergoes constant changes by physical forces.

The figures in Table 1 are obviously the result of calculations involving a number of assumptions. The resulting values are not to be taken as absolute since they involve only an order of probability and a high degree of uncertainty. The percentage column is only relative, and reflects no more than orders of magnitude. Even so, Table 1 does show that the mantle and the core constitute most of the planet, the crust where rational beings live is less than 1%, and the biosphere that contains all life forms is only a very small portion of the crust.

1–2 THE ENERGY SOURCE: THE SUN

The sun is a star 865,000 miles in diameter, about one hundred times the size of earth. Some 5 billion years ago it was a huge distended cloud of dust

Table 1. Composition, Mass, and Volume of the Earth

Region	Composition	Mass ($\times 10^{21}$ metric ton)	Mass (%)	Thickness (km)	Volume ($\times 10^{22}$ km³)
Atmosphere	Nitrogen, oxygen, water, carbon dioxide, inert gases. Gaseous envelope surrounding earth	5	0.00009	15	—
Biosphere	Life area; water, organic (carbon) compounds, skeletal matter; solid, liquid, and colloidal matter	0.0016	0.000003	2	—
Hydrosphere	Salt and fresh water; snow, ice, glaciers, oceans, lakes, rivers, ponds	1410	0.024	3.8 (mean)	137
Crust	Solid silicate rocks, heterogeneous	4300	0.7	30	1,500
Mantle	Silicate materials; iron and magnesium silicates, homogeneous	4,056,000	67.8	2870	89,200
Siderosphere (Core)	Iron–nickel alloy; upper part liquid, lower or central part possibly solid	1,876,000	31.5	3471	17,500
Whole earth	—	5,976,000	100	6371	108,300

and gas that gravitational forces eventually caused to fall inward on itself. The collapse took place over a 50-million-year period, leaving the sun as a sphere of glowing gas weighing about 2×10^{27} tons.

The gas pressure on the sun's core is estimated at 1.3 billion times the air pressure on earth (14.7 lb/in.^{2}), with temperatures of 15,000,000 degrees Kelvin (27,000,000°F). Thermonuclear reactions convert hydrogen to helium, releasing energy.

For about 5 billion years the sun has been about the same size and shape; but in another 5–10 billion years the supply of hydrogen will begin to run low, the core of the sun will reach 100 million degrees Kelvin, and helium by fusion reactions will become carbon, oxygen, and neon. The sun's outer layers will begin to expand and will grow cooler and redder, finally becoming a red giant star. With the onset of fusion reactions producing iron, and further reactions absorbing heat, energy production will slow down and the sun will collapse to a white dwarf star.

The sun is composed of plasma, consisting of atomic nuclei stripped of their electrons, and free electrons. As the degree of collapse increases, the nuclei will be squeezed more closely to a final plasma density of about 1000 tons per square inch, which will counterbalance gravity and stop further collapse.

Then the sun will be a white dwarf about the size of the earth, and about 330,000 times the mass or weight. Over billions of years it will cool off to a cold, dark cinder, circled by frozen planets.

If you are worrying now about the fate of your 100-millionth descendant, consider evolution. Humans may long since have ceased even to look like you, think like you, or act like you, and may have created a world beyond your wildest imagination, perhaps on another planet, or on one of their own making. For those of doomsday persuasion, however, continued worry about the death of the sun will retain its customary attraction.

1–3 PRESSURE AND TEMPERATURE EFFECTS

Theoretically, the earth may be viewed as a smooth, homogeneous, solid sphere, wheeling around the sun, with polarity resulting from the rotation. The ends of the axis differ from other points on the sphere by being relatively at rest. The stars appear to rotate clockwise at the North Pole, and counterclockwise at the South Pole. The poles are reference points by which positions of latitude may be fixed. Rotation regulates the procession of day and night, time, and longitude. The angle of the axis of rotation with the plane on which the earth revolves about the sun determines the position of the Tropics, the Polar Circles, the seasons, the length of daylight and darkness in different latitudes, and the distribution of solar energy on the face of the earth.

This theory, assuming as it does a smooth crust, must be markedly modified to conform with actualities. The crust of the earth, the lithosphere, contains diverse rocks, each with differing thermal properties, specific heats, conductivities, and heating and cooling rates, so that differing temperatures in each result from the same radiation. The crust is not smooth but rugged, with mountains, valleys, and plains—and the oceans, seas, lakes, and rivers (the hydrosphere) cover a very large portion of the crust. Solar radiation falling on the water changes its density and sets up ocean currents between the tropics and the polar areas. The earth's rotation causes a deviation of currents to the right in the Northern hemisphere, and to the left in the Southern hemisphere. The coast lines and inequalities in the sea bed affect the currents, as do submerged mountains. The differential attraction between the sun and moon causes the tidal effects observed everywhere in the hydrosphere.

The earth is enveloped in an atmosphere of air that is never in equilibrium, over land and sea. The atmosphere contains fluctuating quantities of water vapor, and participates in a water circulation cycle whose variance is caused chiefly by solar radiation in the tropics and terrestrial radiation in the polar regions.

Heat is more readily transferred by contact than by direct radiation, and the atmosphere has a more disturbing action over land than over sea. The land develops more friction to the winds, while the sea is smoother. Wind, weather, and frost are eroding agents, along with water on land and in the atmosphere as vapor. Running water, assisted by the detritus (fine rock, sand, and slime) it carries, cuts ravines and valleys and finally spreads the load as alluvium on the plains and as silt on the seashore. Waves driven by the wind may shift the shore or its beaches, or may tumble rocks along the shore line.

The distribution of plant life on land and sea depends upon the temperature and sunlight components of each microclimate. On land, the presence of water is an equal factor, and vegetation is stunted to mosses and shrubs in the cold, moist polar areas. In the temperate zone are the forests, grasslands, and farming areas characteristic of civilization. These are reduced to nothing in the arid tropical areas and brought to profusion in the rain forests of the equator.

The distribution of animals is a function of the availability of plants for animal fodder for the herbivores, and the carnivores who live on them, and this availability is affected by barriers of mountains, deserts, or seas.

Plants and animals have the power of modifying their environment or adapting to it: lagoons become swamps and ultimately meadows by the action of plants; coral reefs are built by animals.

The human animal has overcome the limits set on him by geographical conditions, although there still are aborigines in Australia, stone-age Magasays in the Philippines, pygmies in Africa, and pre-Incan Indians in Brazil's Amazon areas.

THE CRUST (LITHOSPHERE)

2-1 THE STRUCTURE OF THE EARTH

Bruce A. Bolt of the University of California at Berkeley is director of the 16 U.C. seismographic stations. He reported on the structure of the earth's interior, from the study of earthquakes and nuclear explosions.[1]

At the earth's center is a solid inner core with a density about 13.5 times the density of water. The radius of this core is some 1216 km, which makes it a little larger than the moon. The solid inner core is surrounded by a transitional region a little more than 500 km thick. The transitional region, in turn, is surrounded by a liquid outer core about 1700 km thick.

Outside the liquid core is a mantle of solid rock some 2900 km thick, which approaches to within 40 km of the earth's surface under the continents and to within 10 km under the oceans. The thin rocky skin surrounding the mantle is the earth's crust. No drill has penetrated the earth's crust deeper than 8 km.

Figure 1 illustrates the present concept of the earth's structure.

2-2 CONTINENTAL DRIFT AND TECTONICS

Our geologists and oceanographers are continuing to chart the oceans and drill the ocean bottoms; they now present the proposition that all of the lands and sea floors are in constant motion, causing endless changes in the world's structure: the world's continents are adrift on the pliable mantle of inner earth. This process is called global plate tectonics, and there is pictured a turbulent

[1]B. A. Bolt, The fine structure of the earth's interior, *Scientific American, 228*(3): 24–33 (1973).

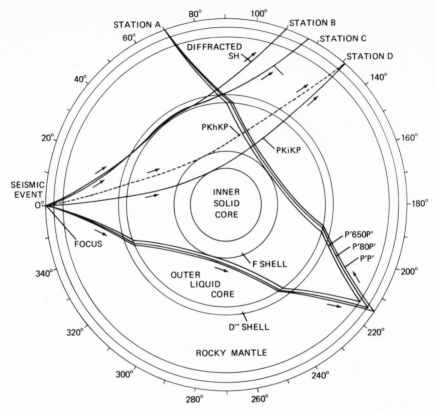

FIGURE 1. Seismic-wave paths and the structural shells of the earth. The three nearly parallel waves arriving at Station A represent wave paths from an underground nuclear explosion. Rays arriving at Stations B and C are diffracted SH and Pc waves, and clarify the nature of the "D" shell. Deeper waves arrive at Station D (PKiKP).

dynamic world where, over a 4.6-billion-year geologic history, oceans have opened and closed like accordians. Continents have been buffeted like hulks on a stormy sea, mountains have formed, ocean beds have been upended, and volcanoes and earthquakes have rumbled.

Far from being a solid, indestructible shell, the earth's crust consists of separate plates (ten major ones, subdivided into varying sizes) made of rock 40–60 miles thick, which float on the hot, viscous mantle beneath. The earth's land surfaces rest on these plates (as do the oceans), and were once together in a single continent. Some 200 million years ago this continent began to split up, eventually forming the seven continents and the major islands. These plates, whose edges are being built up by molten rock welling up from deep fissures in midocean, are being propelled across the globe in various directions at speeds of from one-half inch to six inches per year.

When a moving, land-bearing plate (mainly granite) meets an ocean-bearing plate (consisting of dense, less-buoyant basalt), it rides over it like a bulldozer and scrapes up the sediment deposited on the sea floor over millions of years, along with slices of the crustal rock. The debris piles up along the edge of the land like a rumpled blanket to form mountain ranges. The ocean-bearing plate, forced down at a steep angle into trenches under the land slab, melts from the heat of the friction, forming underground pockets of white-hot lava. The trapped lava is forced up through crevices, erupting on the surface as inland volcanoes. The collision, separation, and shearing of the plates also create earthquakes.

Alfred Wegener, a German scientist, in 1912 theorized that the continents were once joined. He pointed out that the rock formations along the bulge of Brazil and of Africa's Gulf of Guinea are enough alike in age and structure to have been torn from the same geologic flesh, and that identical fossil plants and freshwater animals, which could not have survived a trip across thousands of miles of saltwater, have also been found in South America, Africa, Australia, and even in distant India. No one could conceive of any mechanism that could propel vast continents through the earth's solid crust. Such moving landmasses would have left behind gigantic wakes of displaced rock on the sea floors.

One mystery was the strange lack of sediment on the ocean floors. Sediment formed by microscopic marine organisms and dust blown or washed into the sea should have blanketed the ocean beds over the ages to a depth of at least 12 miles. There is practically no sediment in the center of the Atlantic and only a half-mile veneer near the borders. In the 1850s, engineers laying the transatlantic cable found submerged mountains in midocean. Similar ridges were later found in the Pacific and elsewhere.

In the 1950s, oceanographers discovered that these ridges form a continuous 40,000-mile chain that winds through all of the oceans and that down the center of this chain run deep, hot rifts, oozing lava. The ocean floors were splitting apart and the lava welling up from the ridges was forming new ocean-floor material as it hardened. The crust might be moving farther from the ridge, eventually plunging into the deep troughs bordering the land (Figures 2 and 3).

Geologists have discovered that many times in earth's history the magnetic poles have reversed polarity: during certain ages the iron particles in rock pointed south instead of north. By measuring the extent to which the radioactive elements in the rock have decayed and by determining the age of fossils embedded in them, geologists learned to date and read these magnetic reversals like rings in a tree.

When oceanographers took to the sea in ships, towing magnetometers, they found the magnetic reversal bands as predicted. They were able to determine the age of any particular segment of ocean floor by the direction and speed at which it is moving. They learned, for instance, that the Atlantic

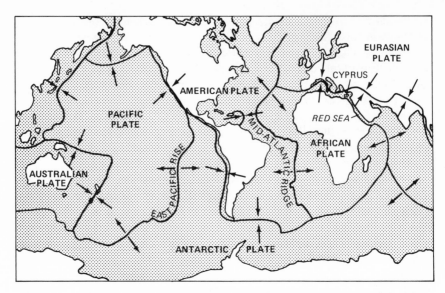

FIGURE 2. The six principal tectonic plates of the lithosphere, the rigid outer shell of the earth. Paired arrows show convergent or divergent plate boundaries.

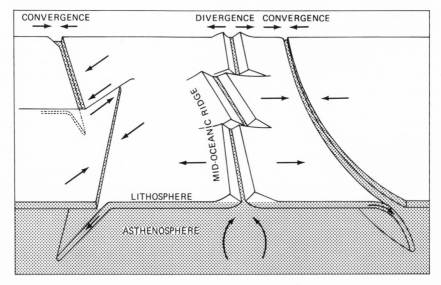

FIGURE 3. The 60-mi-thick lithosphere plates move outward from the divergent plate boundaries of the midocean ridges and plunge convergently downward under the deep-sea trenches.

Ocean floor is widening, pushing Europe and North America apart at the rate of 1 inch per year!

Most geologists agree that the land surfaces were joined 200 million years ago in a single landmass, which they call "Pangaea" (Greek for "all lands"). The heart of this motherland appears to have been located on the equator, in what is now the South Atlantic. Japan lay near the North Pole, India near Antarctica.

The first separation was a gigantic east–west crack in the earth's crust, and a rift appeared between the South American–African mass and that of Antarctica–Australia. India was liberated and started north. It took the plates 200 million years to sever and to reach their present positions.

North America sailed northwest. The Eurasian plate, twisting 20 degrees clockwise, moved north, pursued by Africa, which turned counterclockwise. South America broke loose and headed west, while Greenland and Northern Europe parted. The last two continents to separate from each other were Antarctica and Australia, the latter migrating to warmer climates. The most spectacular of all was India, which, once it tore loose from Africa and Antarctica, traveled 5500 miles north in 180 million years, to ram under the belly of Asia, pushing up the Himalayas ahead of it like a bow wave.

Almost as violent were the happenings in the Mediterranean, where the African plate bashed into Europe. The collision produced the Pyrenees, the Apennines, and the Alps. On the other side of the globe, the Americas, Asia, and Australia plowed toward each other across the vast Pacific. The sea-floor sediment piled up on shore to form the Andes, the ranges of western North America, the island arcs of the Aleutians, Japan, and the other archipelagos of the Western Pacific.

If the plates continue in their present direction, the Atlantic will continue to widen, while the Pacific shrinks. With Africa charging Europe, the Mediterranean seems doomed to become a pond. The Himalayas will grow, and India will tire of burrowing under Asia and will slide eastward. Australia, rocketing northward at 2 inches per year, will sideswipe Asia.

North and South America will still voyage westward, but will part company as Panama and Central America retreat northward. The peninsula of Baja California, and a sliver of the California coast west of the San Andreas Fault, which lie on a separate plate, will tear away from the mainland and head northwest. It will take 10 million years for Los Angeles to sail past San Francisco, and another 50 million years before it slides into the Aleutian trench.

The earth is hardly a stable planet, but civilized man's occupancy is but a brief moment of geologic time. Our viewpoint must be of *now*, if only for a few generations.

About a century ago, the HMS Challenger began a 3½-year voyage over the seas, marking the beginning of oceanic research. This was not long after

Charles Darwin's *The Origin of the Species* appeared. The Challenger studied ocean temperatures, currents, plant and fish life, the flora and fauna of the great ocean basins.

During the debate on plate tectonics, a successor, the Glomar Challenger, began its voyage for the National Science Foundation's Deep Sea Drilling Project. Glomar Challenger is equipped to drill the crust and obtain data to support or deny the geophysical theories of continental drift, sea floor spreading, and the concept of continuously moving crustal plates.

Month after month the Glomar Challenger plies the ocean, drilling sea-floor sediments. It docks every two months for a fresh scientific staff. By 1974, there had been nearly 400 investigators from 25 countries. The cores collected by the ship's drills are transferred to repositories at the Scripps Institution of Oceanography, La Jolla, California, and the Lamont Doherty Geophysical Laboratory of Columbia University in New York. There are more than 25,000 m of recovered cores from 160,000 nautical miles of voyaging, and 450 holes have been drilled in the ocean floor at 300 sites in all the major ocean basins except the Arctic.

In five years of exploring the record of sedimentary sand, silt, volcanic ash, and fossil debris over the ocean crust, there has been confirmation of the relative youthfulness of the ocean crust. Sea-floor spreading and widespread plate movement has occurred, but the pattern of movement is very complex. In addition to the east–west motion away from the midocean ridges where new crust is being formed, the Pacific plate is moving northward as is the floor of the eastern portion of the Indian Ocean. The Glomar Challenger drilling has indicated a major submarine ridge in the Indian Ocean, more than 2000 km long and more than 1 km below sea level. The ridge has deposits of coal. Shallow-water shells indicate a former chain of swampy islands and lagoons. Most oceanic crust subsides with age and moves toward ocean trenches.

The Deep Sea Drilling Project showed petroleum and natural gas in salt domes under the Gulf of Mexico. Sediments may consist of iron and manganese in the lowest layer of sediments just above the igneous crust. Some 5–10 million years ago, crust movement closed the Straits of Gibraltar. Evaporation converted the Mediterranean into a salt-covered desert. The closure was later broken and the Atlantic cut through again to reestablish the Mediterranean over hundreds of years.[2]

Antarctica appears to have been glaciated for at least 20 million years. Drilling has indicated that the ice cap receded 5 million years ago, raising the level of the oceans.

The sedimentary record from drilling shows that the bottom waters of the world's oceans are oxygenated by circulatory patterns that bring cold water

---

[2]K. J. Hsu, When the Mediterranean dried up, *Scientific American*, 227(6): 26–46 (1972).

from the vicinity of the North and South poles to the oceans' depths. An exception is the North Atlantic where prior to 80 million years ago the bottom water was depleted of oxygen. The North Sea is a current producer of petroleum and natural gas. A circumpolar current was established after Australia broke off and separated from Antarctica.

Sedimentary findings indicated the extent and duration of the Antarctic ice cap, how first New Zealand, about 60–80 million years ago, and afterward Australia, moved northward from the south polar regions. Drillings gave evidence that pieces of oceanic crust had been uplifted, tilted, and then subsided. Madagascar, thought to have separated from Africa, was instead a minicontinent for 100 million years. Further study of the cores by interested scientists is expected to reveal much more of the planet's history of the place where we live: the crust.

2–3 THE ANDES AND MOUNTAIN BUILDING

Mountains, valleys, upthrusts, and volcanoes are found all over the world, providing man with outstanding scenery and fulfillment of his aesthetic senses by their overwhelming beauty. The highest mountain peaks on the earth are shown in Table 2.

The Andes have been thought of as young mountains with pinnacles not yet eroded. The Sierra Nevada and the Rockies are estimated to be 100 million years old. The Appalachians (Green Mountains, Adirondacks, Catskills, Poconos, Blue Ridge, the Great Smokies) of the eastern United States, as well as the Laurentians of eastern Canada, are estimated to have been born 450

Table 2. High Mountain Peaks of the World
(*Source: National Geographic Society*)

Table 2a. Highest Peaks in North America

Name	Location	Elevation (ft)
McKinley	Alaska	20,320
Logan	Canada	19,850
Citlaltépetl (Orizaba)	Mexico	18,700
St. Elias	Alaska–Canada	18,008
Popocatepetl	Mexico	17,887
Foraker	Alaska	17,400
Iztaccihuatl	Mexico	17,343
Lucania	Canada	17,147
King	Canada	17,130
Steele	Canada	16,644

Table 2b. Highest Peaks in South America

Name	Location	Elevation (ft)
Aconcagua	Argentina	22,834
Bonete	Argentina	22,546
Sargantay	Peru	22,542
Ojos del Salado	Argentina–Chile	22,539
Tupungato	Argentina–Chile	22,310
Pissis	Argentina	22,241
Mercedario	Argentina	22,211
Huascaran	Peru	22,205
Llullaillaco	Argentina–Chile	22,057
El Libertado	Argentina	22,047

Table 2c. Highest Peaks in Asia

Name	Location	Elevation (ft)
Everest	Nepal–Tibet	29,028
K2 (Godwin-Austen)	Kashmir	28,250
Kanchenjunga	Nepal–Sikkim	28,208
Lhotse 1	Nepal–Tibet	27,923
Makalu 1	Nepal–Tibet	27,824
Lhotse 11	Nepal–Tibet	27,560
Dhaulagiri	Nepal	26,810
Manaslu 1	Nepal	26,760
Cho Oyu	Nepal–Tibet	26,750
Nanga Parbat	Kashmir	26,660

Table 2d. Highest Peaks in Africa, Australia, and Oceania

Name	Location	Elevation (ft)
Kilimanjaro (2 peaks)		
Kibo	Tanzania	19,340
Mawenzi	Tanzania	16,896
Kenya	Kenya	17,058
Margherita	Uganda–Rep. of Congo	16,763
Sukarno	New Guinea	16,500
Pilimsit	New Guinea	15,748
Trikora	New Guinea	15,585
Mandala	New Guinea	15,420
Ras Dashan	Ethiopia	15,158
Meru	Tanzania	14,979
Wilhelm	New Guinea	14,791

Table 2e. Highest Peaks in Antarctica

Name	Elevation (ft)
Vinson Massif	16,860
Tyree	16,290
Shinn	15,750
Gardner	15,375
Epperly	15,100

Table 2f. Highest Peaks in Europe

Name	Elevation (ft)
Caucasus	
El'brus	18,481
Shkara	17,064
Dykh Tau	17,054
Alps	
Mont Blanc	15,771
Monte Rosa	15,203
Dom	14,913
Liskamm	14,852
Weisshorn	14,782
Täschhorn	14,733
Matterhorn	14,690
Dent Blanche	14,293
Nadelhorn	14,196
Grand Combin	14,154

million years ago. They are eroded, rounded off, and in most cases humanized by man in the valleys between the ranges (e.g., the Shenandoah Valley).

The Andean cordillera, or mountain belt, sweeps down from Venezuela, Central America, and Colombia through Ecuador, Peru, and on to the southern tip of Chile. The Peru–Chile trench is very deep, arc-shaped, and roughly parallel to the cordillera, from 4°N latitude to 40°S. From the top of the mountains to the bottom of the watery trench the difference in elevation is 15,000 m, or more than 9 miles. The Andean arc of the mountain ranges and the trench are parts of the "ring of fire" around the Pacific, which is composed of live volcanoes that dot the mountain ranges and of earthquakes associated with ever-present cracks or faults. This circle of fire is completed with the Sierra Nevada and the Rockies in the United States and Canada, the coast of Alaska, Japan, the Philippines, Indonesia, and the Pacific Islands.

Plate tectonic theory[3] holds that the crust of the lithosphere consists of a mosaic of rigid plates in motion relative to each other. The plates are thought, from seismic discontinuities, to be 100 km (62 miles) thick and to include not only the crust but a part of the dense upper mantle. Plate boundaries or junctions seldom coincide with the margins of continents. Magma from volcanoes welling up within the earth's mantle creates the lithospheric plates along ocean ridges. Newly created lithosphere moves away from the ridges to be replaced by magma injected along the axes of the ridges. The spreading plates are consumed at trenches where they bend down and plunge into the earth's mantle.

Earthquakes, volcanoes, land and mountain slides, and mountain building are concentrated along plate junctions, an example of which is the west coast of South America. The oceanic plate, generated along the East Pacific Ridge (Figures 4 and 5), is consumed in the Peru–Chile Trench. The Nazca plate bends down and slides under the South American plate, moving 6 cm per year (2.4 inches per year). The stable continental margin crumples to form belts of folded mountains, the eastern ranges of the Andes, the birth of the Andean volcanic cordillera to the west, and the continental growth of South America.

Mountain peaks are formed by tectonic forces and differential lifting and folding, followed by wearing away of landmasses by the forces of weather, winds, snow, ice, and erosion. The loftiest mountain range in the world is the Himalayas, which culminates in the highest peak in the world, Mount Everest, in Nepal and Tibet. In the Caucasus, in a range in the USSR that divides Europe and Asia, lie the highest peaks of Europe. In western Europe, the Alps, in Switzerland, Italy, and France, have the highest peaks. McKinley in Alaska is the highest peak in North America, and Aconcagua near the border of Argentina and Chile is the highest in South America, while the famed Kilimanjaro is the highest in Africa. More data are given in Tables 2 and 3.

Dr. D. E. James of the Carnegie Institution of Washington is of the opinion that "the Central Andes is a region where mountain-building forces appear to be still at work." The city of Lima, Peru, exists only because engineering skill has developed earthquake-proof foundations and resilient skyscraper construction. The mountains around Lima are huge piles of rounded pebbles larger than ostrich eggs with no smaller sizes that could form a "graded aggregate." North of Lima, the Andes form a single belt of closely spaced mountain chains running parallel to the coast. South of Lima the mountains branch with the easterly fold belt running hundreds of kilometers inland and the westerly volcanic chain continuing parallel to the coast. Between these ranges lies the Altiplano, the area of Bolivia's mineral wealth, and of

[3]J. F. DEWEY, Plate tectonics, *Scientific American*, May (1972).

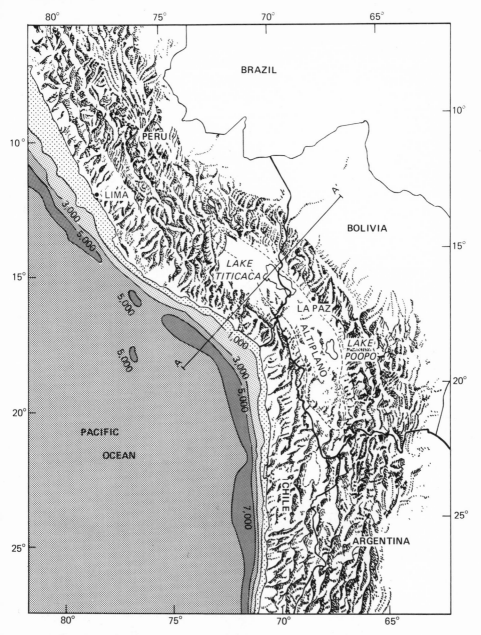

FIGURE 4. Mountain building continues in the Central Andes. North of Lima there is a single belt of parallel, closely spaced mountain chains. South of Lima the mountains branch. The eastern fold belt runs hundreds of kilometers inland; the westerly volcanic group is still parallel to the coast line. The broad, flat plain of the Altiplano lies between the mountain ranges. The plain is underlain by sedimentary debris from the adjacent cordilleras (mountains). The two ranges merge in northern Chile.

FIGURE 5. South America's relationship to the Mid-Atlantic Ridge, the East Pacific Rise, the Pacific Cocos, Nazca, Antarctic, Pacific, the South American plates, and the Peru–Chile trench.

Table 3. Highest U.S. Mountain Peaks

Name	Location (State)	Elevation (ft)
Mt. McKinley	Alaska	20,320
North Peak	Alaska	19,470
Mt. St. Elias	Alaska	18,008
Mt. Foraker	Alaska	17,400
Mt. Bona	Alaska	16,500
Mt. Blackburn	Alaska	16,390
Mt. Sanford	Alaska	16,237
South Buttress	Alaska	15,885
Mt. Vancouver	Alaska	15,700
Mt. Churchill	Alaska	15,638

over 90% of its exports.[4] This broad plain is underlain by a wedge of sedimentary debris eroded from adjacent cordilleras.

Man has built dams and power stations in the Andes which, after initial success in the first few years, were gobbled up by earthquakes.

The atmosphere of Lima appears always to promise rain, but it is an unfulfilled promise. Other than office buildings, Lima homes are often built of adobe (mud) brick which would wash away in even a light downpour.

2-4 EARTHQUAKES

The volcanic structure of the Andes is 15 million years old. At the beginning, huge volumes of volcanic ash, mostly silica, exploded out of fissures and flowed out of cones. These blanketed hundreds of thousands of square meters to a depth of 500 m. About 4 million years ago Andesic lava poured from volcanic vents and formed the great strato volcanoes that are still active.

Only within the crust (lithosphere) are rocks rigid enough to support the brittle fracture of an earthquake.[5] The fracture processes are often initiated by the presence of a notch, which introduces stress, and in rocks these are cracks, openings, or faults in the crust, and are the areas of high strain rate or impact loading, plate movement, and slide displacement. A brittle-failure process proceeds at speeds of fracture approaching several thousand feet per second. The crust shows a complete lack of ductility, little energy absorption, and a brittle or faceted appearance in the fractured surface.

Earthquakes have been known from the beginning of recorded history. In 1858, the term seismology was coined by Robert Mallet,[6] who in his four reports to the British Association catalogued 6831 earthquakes and presented his seismic map of the world. Milne[7] carried the catalog to 1899.

What about those protectors who studied earthquakes to learn more of resistant construction and to predict earth tremors? In Italy, Palmieri constructed his seismograph in 1855, and installed it in the Vesuvius laboratory. Bertelli measured shocks from 1869 to 1878. M. S. DeRossi founded the first journal on earthquakes and volcanoes, the *Bolletino del Vulcanismo Italiano*,

[4]C. L. MANTELL, *Tin*, 3rd ed., Hafner, New York, 1970.
[5]M. F. SHANK, Brittle failure, in *Engineering Materials Handbook*, ed. C. L. Mantell, McGraw-Hill, New York, 1958, Sec. 36-10.
[6]R. MALLET, Catalog of recorded earthquakes 1606 B.C.–A.D. 1841, British Association Reports (1852, 1853, 1854).
[7]J. MILNE, A catalog of destructive earthquakes A.D. 1–1899, British Association Report (1911).

in 1874. Heim and Forel, in 1878, founded the Swiss Seismological Commission which studied earthquakes for three decades and was then merged into a governmental department. Milne, in 1880, founded the Seismological Society of Japan. A system of observers throughout the country aided Milne in cataloging the great earthquakes of Japan. In 1891, the provinces of Mino and Owari had large earthquakes and the Imperial Earthquake Investigation Committee was founded. After the Ischian 1883 earthquake, government activity expanded. In 1895, the Italian Seismological Society started its *Bolletino*. The Laibach earthquake of 1895 helped create the Austrian service. In 1895, the Seismological Committee of the British Association was appointed. After the San Francisco, California, earthquake of 1906, the Seismological Society of America was founded. Field work, observations, and record keeping in the United States is continuously carried forward by the U.S. Geological Survey which monitors all of North and South America.

An earthquake results from a sudden displacement within the crust, a region termed the seismic focus, origin, or center. The vertical extension above, on the surface, is the epicenter. The isoseismal line touches all the points of equal force. The strongest shock is the meisoseismal, or disturbed, area.

Lisbon, Portugal, was destroyed on November 1, 1755, by an earthquake whose epicenter was submarine, west of Lisbon. The disturbed area extended more than a million square miles. The sea retreated, laying bare the sandbar, and then rolled to the equivalent of a 40-foot tide for a two-day period. Inland lakes in Italy, Switzerland, Great Britain, Scotland, Sweden, and Norway were set in oscillatory motion. It is estimated that 50,000 people lost their lives.

In 1783, in southern Calabria, Italy, there were six great earthquakes during February and March, and by 1786 there had been total of 1187 earthquakes. All houses on the nearby plains were destroyed in seconds. Oldham described the 1897 earthquake at Assam, India.[8] The shock was felt over 1.75 million square miles, with serious damage to buildings in an area twice the size of Great Britain. It is believed that the disturbance was at great depth. Fractures occurred in solid rock, fault scarps were widely formed, ridges were displaced, and the ground vibrated like a storm-tossed sea. A previous 1737 earthquake in India killed 300,000 people.

Koto describes the 1891 earthquake in Central Japan (Owari).[9] The 1923 Japanese earthquake had an epicenter beneath Sagami Bay north of Oshima. The mean depth of the focus was 30 miles. Surveys after the disaster showed uplifts of 250 fathoms in one place and depressions of 400 fathoms in another.

[8]R. D. OLDHAM, Report of the Great Earthquake of 1897, *India Geographical Survey Report* No. xxx/x (1899).
[9]B. KOTO, The cause of the great earthquake in Central Japan, *Journal of Colloidal Science Imperial University of Japan*, 5 (1893).

Faulting was clearly evident. The island of Oshima was shifted more than twelve feet east of north and the north shore of Sagami Bay moved about 9 feet east-southeast. Nearly 100,000 people were killed, an even larger number were wounded, and about 44,000 were missing from the earthquake and fires which followed.

Lawson[10] edited the two volumes on the San Francisco, California, and Tarr and Martin[11] described the Alaskan earthquake of 1899.

For a century, seismologists have been developing instruments to forecast earthquakes, mapping dangerous areas, studying earthquake motions, dislocations, faults, aftershocks, secondary effects, distribution frequencies, periodicity, positions of the epicenter, depth of foci, sea waves (often called tidal waves), and the origin of quakes.

It is the hope of seismologists that earthquakes may soon become predictable by study of the geophysical phenomena that precede major spasms. Once precursor signals are discerned, however, the social burden of such knowledge may weigh on us most heavily. Our lives will be little improved if panic and economic chaos result, or if people fail to pay heed to the predictions. The USGS National Center for Earthquake Research[12] is recording and studying the crustal actions for a large region, particularly the Hayward, Calaveras, and San Andreas faults shown in Figure 6. The work will be extended to Southern California, Oregon, Utah, Nevada, and Missouri. Field experiments from a Rangely, Colorado, oil field indicate that small earthquakes may be turned on and off when fluid pressures are varied. A fault may be locked or unlocked by lowering or raising the pressure.[13]

Intensive observations of crustal movements are being made on a 300-km section of the San Andreas fault in Central California. There is a network of 100 seismographic stations and a network of geodetic markers to measure strain along the fault and to detect creep (slow movement under pressure). The seismographs record horizontal as well as vertical motions. More than 4000 small earthquakes were reported in 1972. The network provides the first official earthquake prediction in the United States.

In every earthquake property damage is least on hard rock, more on soft ground, and greatest of all on filled land. Japan has devoted much civil engineering attention to the structural aspects of its architecture. Foundations are wider at the base and taper upward, spreading soil loads. Buildings are

[10]A. C. LAWSON, ed., *The California earthquake of April 16, 1906*, Carnegie Institution, Washington, D.C., 1908–1910, 2 Vol.

[11]R. S. TARR, and L. MARTIN, The earthquakes at Takutat Bay, Alaska, Sept. 1899, *US Geological Survey Professional Paper* 69, Government Printing Office, Washington, D.C. (1912).

[12]*Science 180*(4089): 940–941 (1973).

[13]*Science*, May 25, 1973, p. 851.

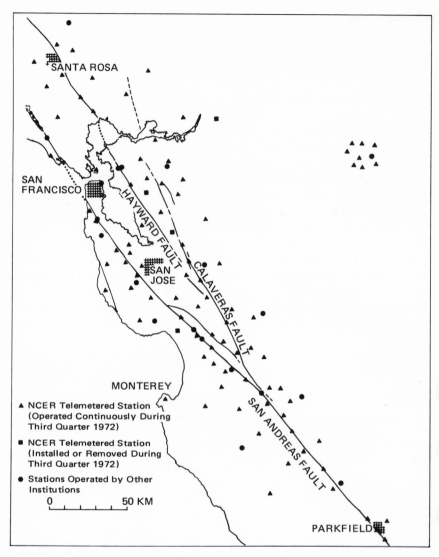

FIGURE 6. Central California, showing faults and seismographs of NCER (National Center for Earthquake Research) and others. (Courtesy, USGS).

framed and braced so that they will move bodily, as one block with its foundations.

In review, Table 4 gives an incomplete list of the major recorded earthquakes for the last 1500 years, with total estimated deaths of 3 billion, or nearly 15 times the present United States population. This toll nearly equals the

3.6 billion people presently on earth, as estimated by Tomas Frejka[14] of the Office of Population Research at Princeton University.

The USGS reported on the geologic and seismologic aspects of the Managua, Nicaragua, earthquake of December, 1972. From 4000 to 6000 people were killed, and 20,000 people were injured. Property damage was more than $0.5 billion. In Managua you can daily view active volcanoes. The "black" waters of Lago Nicaragua not only cannot be swum because there are fresh-water sharks, but because it receives all the waste of the city. The city is situated directly on four active faults, each of which moved during the earthquake. Geologic hazards make this an area of unusually high risk. Historic records indicate previous earthquakes and prior episodes of fault movement. Managua had a devastating earthquake in 1931. Historically, the memory of the human race is short.

Geologists can become protectors of us all. They are providing advice on how to rebuild Managua to withstand geologic hazards, and to resist earthquake damage in its reconstruction. And although they may say "Don't live there, it's dangerous," they seriously wonder whether anyone will listen to them.

"The city that waits to die" was a British Broadcasting Corporation broadcast of the high probability of a major earthquake in the San Francisco area of California. San Fernando had an earthquake in 1971, when the coastal block of the earth's crust on which Los Angeles sits moved northward and under the San Gabriel Mountains which were lifted as high as six feet in some sections. In 1972, the Oxnard quake between Los Angeles and Santa Barbara started about 12 miles off the coast and acted in the manner of the 1971 San Fernando movement (Figure 7). There is motion that pushes Central California west, relative to Southern California, generating the Garlock Fault and producing a bend in the San Andreas Fault. It leads to a build-up of pressure, producing underthrusts, and may help set the stage for great earthquakes such as that in San Francisco in 1906.

A whole series of studies by commissions, including the Federal Office of Emergency Preparedness, the USDI, the USGS, the Weather Bureau, and assorted insurance research teams have assumed the function of protectors. As a result the city is better prepared than it was at the time of the 1906 earthquake, when 700 persons were killed. The high-rise buildings where fires may break out are expected to remain standing.

The San Francisco study leaned heavily on the observed effects of the most severe shaking in the San Fernando earthquake and sought to take into account such factors as time of day (whether people are home in bed, commuting, or at work) and season (whether rain-soaked slopes are likely to slide).

The worst combination of factors—an earthquake like that of 1906, occurring at 4:30 P.M. along the San Andreas Fault near the city—could take

[14]T. FREJKA, *Scientific American*, 228(3): 15–23 (1973).

Table 4. Major Recorded Earthquakes Throughout the World

Date	Location	Deaths	Comments
5/20/526	Antioch, Syria	250,000	Destroyed city
7/9/551	Beirut, Syria	—	Leveled town
12/1/856	Corinth, Greece	45,000	Leveled city
936	Constantinople	—	
1057	Chihli, China	25,000	Destroyed area
1268	Silicia, Asia Minor	60,000	Destroyed area
9/27/1290	Chihli, China	100,000	Destroyed area
5/20/1293	Kamakura, Japan	30,000	Destroyed area
1/26/1531	Lisbon, Portugal	30,000	Followed by fire
1/24/1556	Shensi, China	830,000	Wiped out a civilization
11/1/1667	Shemakha, Azerbaijan	80,000	Wiped out a civilization
1/11/1693	Catania, Italy	60,000	Destroyed city
12/30/1703	Tokyo, Japan	200,000	Destruction and sea waves
10/11/1737	Calcutta, India	300,000	Ground opening and fire
6/7/1755	Northern Persia	40,000	
11/1/1755	Lisbon, Portugal	60,000	Second destruction of city in 25 years
2/4/1783	Southern Italy and Sicily	50,000	Messina ravaged
2/4/1797	Cuzco, Peru; Quito, Ecuador	40,000	Both cities destroyed
9/5/1822	Aleppo, Syria	22,000	Heavy destruction of men and animals
12/28/1828	Echigo, Japan	30,000	
8/13–15/1868	Peru and Ecuador	25,000	Property damage: $300 million
5/16/1875	Venezuela and Colombia	16,000	Destruction of city
4/22/1884	Colchester, England	—	
6/15/1896	Senriku Coast, Japan	22,000	Seismic sea wave
4/18/1906	San Francisco, California	700	Followed by fire: $500 million loss

Date	Location	Deaths	Notes
8/16/1906	Chile	1,500	$100 million loss
12/28/1908	Southern Italy and Sicily	7,500	Wide area destruction
1/13/1915	Central Italy	30,000	Wide area destruction
12/16/1920	Kansu, China	180,000	Ten cities demolished
9/1/1923	Yokohama and Tokyo, Japan	143,000	All of Yokohama and half of Tokyo obliterated
12/26/1932	Kansu, China	70,000	Second earthquake in 12 years
5/31/1935	Quetta, Baluchistan, and India	60,000	Also earth landslides
1/24/1939	Chile	30,000	
12/27/1939	Anatolia, Turkey	23,000	Series of earthquakes followed by floods
12/21/1946	Southern Japan	2,000	Followed by six seismic waves
6/28/1948	Fukiji, Japan	5,131	Quake and fire destroyed most of Fukiji
8/5/1949	Ecuador	6,000	Heavy damage to 50 towns
6/10–17/1956	Northern Afghanistan	2,600	Continued series of quakes
7/2/1957	Iran	2,500	Tremors along shores of Caspian Sea
12/2/1957	Outer Mongolia	1,200	
12/13/1957	Western Iran	2,000	
2/29/1960	Agadir, Morocco	12,000	Followed by seismic wave and fire
5/21–30/1960	Chile	5,700	Followed by seismic wave and fire
1/10/1962	Peru	2,000	Avalanche on Hisascaran, extinct volcano
9/1/1962	Northern Iran	10,000	
8/19/1966	Turkey	2,530	100,000 homeless
8/15/1968	Donzgala, Indonesia	200	
8/31/1968	Northeast Iran	13,000	World's worst earthquake in 30 years
3/28/1970	Western Turkey	1,089	
5/31/1970	Northern Peru	50,000	Avalanche followed
4/10/1972	Southwestern Iran	5,374	
12/23/1972	Managua, Nicaragua	10,000	Capital city destroyed by earthquake a second time

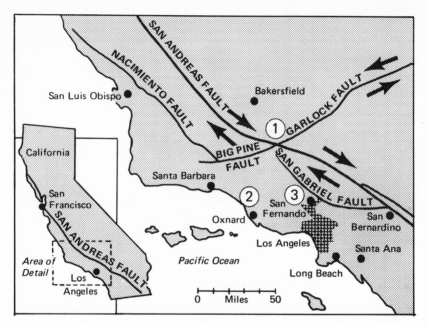

FIGURE 7. A bend (1) in the San Andreas fault where it crosses the Garlock Fault impedes the northwest slippage of California's coastal rim along the San Andreas. (2) denotes the 1974 Oxnard, and (3) the 1971 San Fernando earthquakes. Further north, slippage near San Francisco is also blocked, allowing stress build-up for a major earthquake.

an estimated 10,360 lives and could produce 40,360 casualties requiring hospitalization. However, if such an earthquake were at 2:30 A.M., the casualties would be 2300 dead and 10,800 injured.

The Golden Gate and Bay bridges leading into San Francisco were built to withstand earthquakes, and the chief concern is for approaches such as the one to the Bay Bridge in Oakland where landfill has been laid over bay mud.

The airports, their runways, and their structures are built on bay mud, unstable in earthquakes; when one comes, the airports would be wrecked. So would electric power, gas lines, and communication. An expected fire would leave 20,000 people homeless, as compared to 225,000 in 1906.

The present water system has been designed to isolate major breaks, and fire-fighting techniques have been built up, along with disaster hospital units to prepare for the expected and, to the protectors, inevitable earthquake. San Francisco expects to ward off the punches and survive.

In 1973, Mexico was hit by a massive earthquake, described as its worst in decades. More than 425 persons were killed and additional hundreds were injured as the earthquake ripped through the country's storm-ravaged central section. Thousands were added to those already left homeless by earlier flooding. The earthquake affected a three-state area south and west of Mexico City. The earthquake measured variously at 5.5–6.5 on the Richter scale. Any reading of 4.5 or above is considered potentially dangerous. Electric power and telephones were knocked out in many areas, slowing rescuers and delaying casualty reports. The earthquake hit Mexico when it was already suffering from severe floods and a hurricane that had killed 70 people and damaged the homes of 400,000 others, in the months of July and August, 1973.

Established cultures often introduce hazards to man's work. The oldest city on the American continent is Mexico City, almost as big as New York. Mexico City, Distrito Federal, is the capital and rests on the bed of a drained lake, Texcoco. The Aztecs settled on an island in the lake in 1325. By a system of drainage, dikes, and causeways, they expanded the capital of an empire that extended over most of central and southern Mexico. In the center they built high pyramids and temples to their gods to whom they offered human sacrifices. Hernan Cortes, the Spaniard, conquered the city in 1521, razed the pagan buildings, introduced Catholicism, and built anew on the ruins.

The Aztecs' choice of location has resulted in a severe problem for modern Mexico City. It shows dramatically at the Basilica of Our Lady of Guadalupe. The west side, with its bell towers, stands straight; the east towers and the chapel tilt crazily. The west side rests on a solid foundation of rock, the east side on the mud of Lake Texcoco. Some areas underlaid with mud have sunk as much as 25 ft since 1900.

The subsoil is mostly volcanic ash and structured somewhat like a honeycomb—but the honeycomb is filled with water. The Mexicans drew on the water in the honeycomb for their water supplies, and the volcanic ash particles consolidated. Water now comes from other sources, and the rate of sinking has been slowed, but between 1948 and 1950, subsidence occurred at the rate of 31 inches per year in places. Sewer lines, instead of flowing downward out of the city, tilted up and the flow reversed. If the pumps ever failed, the inhabitants would be swimming in sewage. Tunnels are now being built between 160 and 720 ft below the surface, to drain the sewer system under the mountains into an arid area north of the city.

The 43-story Latin American Tower is anchored by 360 steel and concrete piles 108 ft deep, and these support half the building's weight. The rest of the load floats on a box- or boat-like foundation 45 ft below the surface. The water content in the soil is kept constant by automatically injecting more water whenever the pressure drops. The tower has not sunk at all since its completion 17 years ago, and in 1957 it escaped unscathed from the severest earthquake in Mexican history, which damaged 975 buildings in the city. Every building around without similar construction gradually sinks and tilts.

On Sundays and holidays, with factories shut down, there are times when there is a view of the 17,887-foot-high volcano Popocatepetl, and its companion, Iztaccihuatl, southeast of the city. Mexico City is subject to earthquakes, dirty air, and smog. However, there are also the Alameda and Chapultepec parks, color, crowds, shantytowns, laughter, and affection.

2-5 VOLCANOES

Volcanoes are openings, cracks, or pipes in the earth's crust, through which molten rock or magma from the asthenosphere (the plastic layer beneath the crust and around the core) may exude, erupt, or blow out to relieve crustal pressures. In the language of the physicians, volcanoes might be hernias, and in the language of chemical engineers, they might be ruptured disks, or failed seals on very high pressure autoclaves.

The volcano forms a hill, conical in shape, usually with a hollow, or crater, on top. All the attendant phenomena are studied under the term vulcanology, after the Roman deity, Vulcan, lord of fire and metal working.

An eruption is usually preceded by earth movements, earthquakes, subterranean sounds, changes in temperature, the flow of springs, and the evolution of gases near the crater. Emissions of vapor, often attaining tremendous volumes, occur from the beginning to the end. Pliny, the Greek historian, described this for the Vesuvius eruption in A.D. 79. The steam and dust of Vesuvius in April, 1906, arose to a height of 6–8 miles. The fall of ash and cinders crushed the roofs of buildings and burned the villages of Ottajano and San Gisiseppe. The Kilauea volcano in Hawaii has disgorged more than 440 million cubic yards of lava since May, 1969.

The steam from erupting volcanoes condenses to rain which mixes with ashes and loose material to form mud that flows down the cone. Such mud buried Herculaneum in a flood from Vesuvius in A.D. 79, sealing the city. Subsequent eruptions covered the area with lava. A torrent of mud was the earliest phenomenon from the eruption of the Mount Peleé volcano in Martinique in 1902.

When a volcano ceases to be active, the throat plugs with cooled rock. When it again becomes active, large blocks may be ejected.[15] The volcanic cinders, sand, ashes, and dust are forms of lava, as is "Peleé's hair" (the Hawaiian god of volcanoes), a filamentary material resembling the synthetic product, rock wool. Sometimes the lava cools rapidly to form a volcanic glass, obsidian, often used as a semiprecious gem. There are glass mountains of black obsidian in the Brazilian Eastern Sierras, preserved as a national park. Glassy lava may be ejected in a light cellular form, or sponge, such as pumice.

[15]C. L. MANTELL, *Engineering Materials Handbook*, McGraw-Hill, New York, 1958, Sec. 25, p. 20.

Vapors and gases exist in rock and rock layers (magmatic), and are expelled when the rock is melted. Most of the gases are water vapor or carbon dioxide. Vapor vents are often termed fumaroles, and they yield mixtures of water and volcanic ash, mostly mud, and are related to geysers. Mud volcanoes occur in Iceland and Sicily. In some volcanoes, sulfur is formed by the reaction of hydrogen sulfide and sulfur dioxide. Ammonium chloride and salt, as well as other chlorides, are deposited. Sulfur has been mined and recovered in Sicily for many decades.

Shepherd reported on the gases evolved at the lava lake in Kilauea, Hawaii.[16] From analyses of this and other volcanoes, it appears that water, carbon dioxide, sulfur, sulfur dioxide, hydrochloric acid, and ammonium chloride are often abundant.

Zies reported that the fumaroles in the Valley of Ten Thousand Smokes in Western Canada gave off water (98.8–99.9%), carbon dioxide, carbon monoxide, oxygen, methane, hydrogen sulfide, nitrogen, hydrogen, argon, and hydrochloric and hydrofluoric acids. The hydrochloric acid was only 0.117% and the hydrofluoric acid 0.032%. The total quantities of those acids produced in a year were, quite remarkably, 1,250,000 tons and 200,000 tons, respectively. Those readers who are concerned about clean air might compare these figures with the smaller United States and Canadian production of hydrochloric and hydrofluoric acids.

In a large number of cases, a volcanic crater is a small circular hollow around the orifice, or opening. In others, there is a large bowl-like cavity, or caldera. In Hawaii, there are wide pits and vertical walls, floored by a great plain of black basalt, or with lakes of lava. Perhaps the largest crater is that of Asosan, in Kyushu, Japan, with an area of 100 square miles.

Submarine volcanoes associated with the mountain ranges of the Atlantic and Pacific floors have been predicted, observed, and located. The surface of the sea is violently agitated, and the water boils with huge fountains, shoals of dead fish, cinders, bombs, pumice, and rock. A little island appears, the peak of a cone based on the ocean floor. Christmas Island in the Indian Ocean is in water more than 14,000 ft deep. Many volcanic islands are spread over the Pacific.

Sometimes volcanic plugs are created which do not break through the crust but leave pipes of "blue earth," which are diamond bearing. Diamonds are a high-temperature, high-pressure form of carbon. Mantell has discussed the origin, mining, and geology of diamond mines in Africa and South America.[17]

Volcanoes are most frequent in the Pacific circle of fire, mentioned in connection with earthquakes. Volcanoes range along lines of fracture, particularly at intersections of two or more fracture lines, where the crust is thinned or

[16]L. SHEPHERD, *American Journal of Science*, *235A*: 321 (1938).
[17]C. L. MANTELL, *Carbon and Graphite Handbook*, J. Wiley & Sons, New York, 1968.

weakened. The Chilean, Bolivian, and Peruvian Andes are studded with extinct, dormant, and active volcanoes, including the highest in the world. Ecuador has Cotosaxi, 19,600 ft high, and Sangay, which has been regarded as one of the most active. In Managua, Nicaragua, active volcanoes are visible from the city. The Plateau of Mexico has volcanoes in a band of country from Colima to Tuxtla near Vera Cruz, including Orizaba, 18,200 ft high, and Popocatepetl, 17,880 ft high.

The United States has very few active volcanoes, although many became extinct only in recent geologic times. Très Virgines erupted in 1857, and Lassen's Peak occasionally smokes. The Moon Valley craters and Mount Shasta are extinct. The Cascade Range contains volcanic peaks, and both Mt. Hood, in Oregon, and Mt. Rainier, in Washington, exhale vapor. Mount St. Helena, in Washington, erupted in 1841 and 1842, and Mt. Baker was active in 1843.

Alaska has volcanic activity along the Coast Range and in the adjacent islands, such as Mt. Fairweather, Mt. Wrangell on Copper River (erupted in 1819), Cook's Inlet, St. Augustine, and two volcanoes on Unimak Island. The Aleutian volcanic belt stretches 1600 miles around the Pacific.

Kamchatka Peninsula has 14 volcanoes, and the belt continues to the Kurile Islands and to the Japanese archipelago, where 54 volcanoes are active or recently extinct; the best known is the sacred Mt. Fuji. Bandai-San, north of Tokyo, erupted in 1888, when it blew off one side of the peak, Kobandai, removing nearly 3 billion tons of material.

South of Japan, the train of volcanoes continues through the Philippines, Taiwan, and the Moluccas and Sunda archipelago, where almost every island teems with volcanoes, solfataras (named after La Solfatura, a dying volcano near Pozzuoli, Italy; a fumarole that yields hot vapor and sulfurous gases), fumaroles (smoke holes or vents that issue volcanic vapors), and hot springs, through New Guinea, the Pacific Islands of the Marshall group, and New Zealand. Tarawera was in eruption in 1886. When the three-peaked mountain suddenly erupted, a huge rift opened and Lake Rotomahana subsided. South of South Victoria Land in the Antarctic are the volcanoes Erebus and Terror.

Within the basin of the Pacific there are a vast number of volcanic islands rising from deep waters and having active craters. In the Hawaiian group there are some 15 large volcanic mountains. Mauna Loa and Kilauea are active. Hualalai last erupted in 1816.

The volcanic regions of the Pacific are connected to those of the Indian Ocean through the islands of Indonesia, the Moluccas, Sunda, Java, and Sumatra; they form a system of crust fissures. Volcanic peaks occur in Flores, Sumbawa, Lampoc, and Bali. A terrific eruption at Tambora, in 1815, was one of the largest recorded. Java contains 50 or more great volcanic mountains, active, dormant, or extinct. Papandayang erupted with great violence in 1772.

Krakatau, in the Strait of Sunda, was once a volcano of gigantic size. In 1883, after 200 years of repose, part of the island was blown away. The column of dust, cinders, ashes, and steam was 20 miles high and the dust traveled around the world. Vast quantities of pumice were carried hundreds of miles by sea currents before they became waterlogged and sank.

Dr. W. T. Pecora, then director of the U.S. Geological Survey, stated that the particulate matter dispersed from volcanoes is a continuing phenomenon. From eruptions such as Krakatau and Mount Kamai, in Alaska, in 1912, and Hekla, in Iceland, in 1947, more particulate dust and ash were ejected into the atmosphere than from all of man's activities and from his cities, factories, fires, or other works. Man's presence often seems truly insignificant among the volcanoes from which he must so often retreat.

The chain of volcanoes continues northward through Sumatra to the Bay of Bengal (Barren Island, active; and Narcondam, extinct), through the Mascarene Islands, and active craters in Reunion and far south in the Indian Ocean, and the volcanic islands of New Amsterdam, St. Paul, and the Comoros in the channel of Mozambique.

In Africa, there are a number of extinct volcanoes, but some partly active ones are associated with the Rift Valley and with large plains areas which formerly were volcanic caldera and are now wild animal conservation districts. The enormous cones of Kenya and Kilamanjaro have craters. The Mfumbiros, south of Lake Edward, are 14,000 ft high. Kirunga, north of Lake Kivu, is still partly active. Elgon is an old volcanic peak. On the west side of Africa, Cameroon Peak was active in 1909. The island of Fernando Po is volcanic, as are the Jebel Teir, the Twelve Apostles Islands. Aden is located on the wreck of an old volcano.

In the Atlantic, a broken band of volcanoes is traceable from the submarine ridge. Jan Mayen, in the Arctic, erupted in 1818. Iceland has about 130 postglacial volcanoes, and about 30 have erupted in man's history, the latest being Surtsey, in 1973, fortunately with no loss of life despite great destruction of villages and the fishing industry. Iceland had an immense fissure flow of lava from Laki, in 1783, through cracks in the earth's crust. Fissure flows were responsible for sheets of old lava in the terraced hills of the Faroe Islands, the west of Scotland, and the north of Ireland. The Azores are a volcanic archipelago, as are the Canaries, the Cape Verde Islands, Ascension, St. Helena, and Tristan da Cunhá.

In the West Indies, there are many volcanoes, solfataras, and hot springs associated with a huge earthfold and deep water. The catastrophic outburst of Mt. Pelée in Martinique, in 1902, wiped out the city of St. Pierre, causing 40,000 deaths. There was a sudden emission of a dense cloud of superheated and suffocating gases, heavily charged with incandescent dust, moving with great velocity, and accompanied by the discharge of immense volumes of volcanic sand. The sand descended like a hot avalanche. The cloud at Pelée was visible from islands many miles away as a solid bank, opaque and impenetrable. It produced an intense darkness, with flashes of lightning.

On the mainland of Europe, Vesuvius is still active and has been the subject of numerous studies, as well as of historical novels and plays. In the Mediterranean, there is Etna on the Sicilian coast, and Stromboli and Vulcano are chronically active in the Lipari Islands. There have been many submarine volcanoes in the Mediterranean since Grecian times. The remains of many extinct volcanoes can still be seen in Auvergne, Eifel, Bohemia, Catalonia, and Italy.

Since the earliest recorded history, there have been many thousands of volcanic disasters. We have only occasionally been able to help the victims, and have never been able to change the vulcanism itself. The major known volcanic eruptions are listed in Table 5. Their casualties number less than 200,000 people, which is small compared with earthquakes, perhaps because volcanoes often give warnings in advance. The great volcanoes of the world are summarized in Table 6.

More than 500,000 square miles of the Earth's surface have been scarred, within historical time, by molten lava, rocks, cinders, ash, or gases expelled by volcanoes. Although many volcanoes can be recognized by their graceful cone-shaped mountains—such as the revered Fuji of Japan—others lack prominent cones and are typified by sluggish lava flows emerging from long cracks, or fissures, as in Hawaii.

About 600 active and 10,000 inactive, or dormant, volcanoes are found along three belts, closely matching the earthquake regions of the world. The circum-Pacific belt, or "ring of fire," includes the Pacific coasts of North and South America as well as Asia. The second belt extends from the Mediterranean to the East Indies. The third lies along the crest of the world's longest mountain chain—the midocean ridge connecting the major oceans.

Volcanic activity is deceptively sporadic: brief periods of outgassing, or eruptions, followed by unpredictably longer dormant periods. History reveals that many societies have been devastated by presumably extinct volcanoes.

Archeologists feel that the Greek island of Thera, from excavations made there in 1966–67, was the fabled Atlantis, with a population of 30,000. Thera is in the Aegean Sea 70 miles north of Crete and was the site of an ancient Minoan city destroyed by a volcano in 1500 B.C.

Cnossus, on the northern coast of Crete, was settled before 3000 B.C. and was the center of the Minoan culture that was eventually to be partly absorbed by European culture. An earthquake circa 1500 B.C. destroyed the palaces of King Minos, but, nothing learned, the city was rebuilt on the same site.

Pompeii was a resort for wealthy Romans on the Tyrrhenian Sea (near modern Naples). It was destroyed by an earthquake, along with Herculaneum

Table 5. Major Volcanic Eruptions
(*Courtesy National Geographic Society*)

Date	Location	Deaths	Comments
8/24-25/79	Pompeii, Herculaneum, and Stabiae, Italy	16,000	Mount Vesuvius erupted; destroyed these towns
1169	Sicily	15,000	Mount Etna erupted
12/16/1631	Italy	4,000	Mount Vesuvius erupted and destroyed five towns
1772	Java	3,000	Mount Papandayang erupted
1783	Iceland	9,000	Mount Hekla erupted; 20 villages obliterated by lava, and many more flooded by water
1815	Java	12,000	Tambora exploded, followed by violent whirlwinds and tidal waves
8/26-28/1883	East Indies	35,000	Krakatau erupted; most of the island destroyed; Java and Sumatra heavily damaged by tidal waves
4/8/1902	Guatemala	1,000	Santa Maria erupted
5/8/1902	Martinique, West Indies	40,000	Mount Pelee erupted, wiping out the city of St. Pierre
1911	Luzon, Philippines	1,400	Mount Taal erupted
1919	Java	5,000	Mount Kelud erupted; 100 villages destroyed
1/18-21/1951	New Guinea	3,000	Mount Lamington erupted
4/28/1966	Java	1,000	Mount Kelud erupted; destroyed nine villages
1/23/1973	Heimay, Iceland	—	Kirkjufell Volcano erupts; 500 homeless

Table 6. Great Volcanoes of the World
(Courtesy National Geographic Society)

Continent	Name	Location	Height (ft)	Status
Africa	Kibo, (Kilimanjaro)	Tanzania	19,340	Dormant
	Cameroon Mtn	Cameroons	13,350	Erupted in 1959
	Nyiragongo	Rep. of the Congo	11,385	Erupted in 1948
	Nyamagira	Rep. of the Congo	10,028	Erupted in 1958
Antartica	Erebus	Antarctica	12,450	Steaming
Asia	Klyuchevskaya	USSR	15,584	Erupted in 1962
	Kerinchi	Indonesia	12,467	Steaming
	Fuji	Japan	12,388	Dormant
	Rinjani	Indonesia	12,224	Erupted in 1964
	Tolbachik	USSR	12,080	Erupted in 1941
Mid-Pacific	Mauna Loa	Hawaii	13,680	Erupted in 1950
	Kilauea	Hawaii	4,090	Erupted in 1969
Europe	Etna	Sicily, Italy	10,958	Erupted in 1969
North America	Popocatepetl	Mexico	17,887	Steaming
	Wrangell	Alaska	14,005	Steaming
	Colima	Mexico	13,993	Steaming
	Spurr	Alaska	11,069	Erupted in 1953
	Torbert	Alaska	10,600	Erupted in 1953
Central America	Tajumulco	Guatemala	13,812	Rumbling
	Tacaná	Guatemala	13,333	Rumbling
	Acatenango	Guatemala	12,992	Rumbling
	Fuego	Guatemala	12,582	Erupted in 1967
	Santa Maria	Guatemala	12,362	Rumbling
South America	Guallatiri	Chile	19,882	Erupted in 1959
	Lascar	Chile	19,652	Erupted in 1951
	Cotopaxi	Ecuador	19,347	Steaming
	Misti	Peru	19,031	Dormant
	Cayambe	Ecuador	18,996	Dormant

and Stabiae, in A.D. 63, but was rebuilt. In A.D. 79, Mount Vesuvius, 1 mile away from the three cities, erupted and buried all three cities under volcanic ash, mud, and stones. Pompeii was accidentally discovered in 1748, excavated, and slowly restored. Murals and statues were well preserved by the lava.[18]

Three of the seven ancient wonders of the world were destroyed by earthquakes. The Colossus of Rhodes was a massive, 105-ft-high bronze statue of the sun god Helios, located on a promontory overlooking the harbor of Rhodes. It was toppled by an earthquake in 244 B.C. The Pharos, or Lighthouse, at Alexandria, was built around 280 B.C. by Ptolemy II, Greek King of Egypt. It was 600 ft high and its beacon was visible for more than 40 miles. It was demolished by an earthquake in the fourteenth century. The tomb of Mausolus of Crete was erected by his wife around 352 B.C. at Halicarnassus, south of Izmir, Turkey, a white marble structure of 36 Ionic columns, and was the work of two Greek sculptors, Scopas and Praxiteles. It was destroyed by earthquakes before the fifteenth century and its marble has since been reused.

The earth's crust is not stable, but in dynamic motion, its forces following purely physical laws that are far beyond our anticipation or control, or even, usually, our ability to recognize the inevitability of natural processes. As the earth cools, the crust on which we exist will stabilize, but that will occur only in geologic time, over a span of millions of generations.

The Valley of Ten Thousand Smokes flamed into being in 1912, when Mount Katmai in the Aleutian Range of western Alaska erupted. Mt. Katmai lost 2500 ft of its cone, or head, and the outpouring left a crater 8 miles in circumference and two-thirds of a mile deep. One thousand square miles of surroundings, including Kodiak Island, were buried in pumice and ash. The forested valley north of Mt. Katmai was turned into a vast lake of bubbling incandescent rock as explosions tore massive holes in the permafrost of the earth. The area was denuded of trees. By 1916, the valley had become a seething mass of fumaroles and mud pots, and emitted clouds of steam and gases, including hydrochloric and hydrofluoric acid, as well as water, carbon monoxide, and carbon dioxide.

When Robert Griggs saw the 20-mile valley in 1916, there were thousands of geysers hundreds of feet high. The fuming continued until the 1940s, when a calmer period set in. In 1916, President Wilson created the Katmai National Monument, including the Valley of Ten Thousand Smokes. Twice enlarged, it now covers 4363 square miles and its features include charcoal stumps, a pink desert, numberless circles of red and purple ex-fumaroles, orange and yellow gas vents, arctic flowers, and Alaskan brown bears.

---

[18]SMITHSONIAN, Washington, D.C., Oct. 1972, Spectacular new finds made in inexhaustible Pompeii.

2-6 COMPOSITION OF THE CRUST

B. Mason[19] states that the average composition of the crust is in effect that of the igneous rocks, since the amount of sedimentary and metamorphic rocks is small. Clark and Washington[20] estimated an average composition of the crust, on a dry basis, with water of composition omitted (Table 7).

Clark and Washington estimated the upper portion of the crust to be 95% igneous rock, 4% shale, 0.75% sandstone, and 0.25% limestone. Sedimentary rocks form a thin veneer on an igneous rock foundation.

Table 8 shows an estimated amount of the chemical elements in the earth's crust from the viewpoint of the geochemist.

Eight elements (oxygen, silicon, aluminum, iron, calcium, sodium, potassium, and magnesium) make up 98.5% of the earth's crust, and, with titanium, hydrogen, phosphorus, and manganese, 99.3%. Oxygen in bound form constitutes nearly half, present as silicates of aluminum, calcium, magnesium, sodium, potassium, and iron. Only three of these exist as structural elements: aluminum, iron, and magnesium, with manganese as an alloying element for these.[21]

The elements important to our physical, protective, and economic life are very small in amount. Copper is less abundant than zirconium, lead is less abundant than lanthanum or tin, zinc is less abundant than vanadium or chromium, tin is less abundant than cerium, while cobalt and molybdenum, as well as tantalum and mercury, are rarer than the so-called "rare elements."

These figures mean very little when the heterogeneous nature of metal deposits, their concentration by physical forces, natural corrosion, erosion, hydraulic flows of streams, earth movements, thermal pressures, volcanic forces, wind, weather, and the water cycle are all considered. The metals are found in metallogenetic provinces: for example, iron is found in the Cayuna Range in Minnesota; lead is found in Missouri; zinc and lead are found in the tri-state area of Missouri, Arkansas, and Oklahoma, as well as in Tennessee and Kentucky; magnesium is derived from seawater; nickel is found in Sudbury, Ontario; cobalt and silver are found in Cobalt, Ontario; and copper is found in Montana and Arizona.[22]

Metals produced by man are iron and steel, measured in units of 100 million tons annually; aluminum, in units of 10 million tons; copper, lead, and zinc in units of millions of tons; tin, nickel, and magnesium in units of 100,000 tons; and calcium and lithium, as metals, in units of 100 tons.

[19]B. Mason, *Principles of Geochemistry*, J. Wiley & Sons, New York, 1952.
[20]F. W. Clark, and H. S. Washington, The composition of the earth's crust, *US Geological Survey, Professional Paper* 127, 1924, 117 pp.
[21]C. L. Mantell, *Engineering Materials Handbook*, McGraw Hill, New York, 1958, 2000 pp.
[22]C. L. Mantell, *Engineering Materials Handbook*, McGraw-Hill, New York, 1958, Sec. 23, pp. 2-18.

Table 7. Estimated Average Dry Composition of the Crust

Compound	Percentage
Silica	60.2
Alumina	15.6
Ferric oxide	3.1
Ferrous oxide	3.9
Iron oxides	7.9
Magnesia	3.7
Lime	5.2
Sodium oxide	3.7
Potassium oxide	3.2
Titanium dioxide	1.1
Phosphorus oxide	0.3

Table 8. Estimated Concentration of Elements in the Crust

Element	Symbol	Parts per million
Oxygen	O_2	446,000
Silicon	Si	227,200
Aluminum	Al	81,300
Iron	Fo	50,000
Calcium	Ca	36,300
Sodium	Na	28,300
Potassium	K	25,900
Magnesium	Mg	20,900
Titanium	Ti	4,400
Hydrogen	H	1,400
Phosphorus	P	1,180
Manganese	Mn	1,000
Sulfur	S	520
Carbon	C	320
Chlorine	Cl	314
Rubidium	Rb	310
Fluorine	F	300
Strontium	Sr	300
Barium	Ba	250
Zirconium	Zr	220
Chromium	Cr	200
Vanadium	V	150
Zinc	Zn	132
Nickel	Ni	80
Copper	Cu	70
Tungsten	W	69
Lithium	Li	65
Nitrogen	N_2	46

2-7 THE ROCKS WE USE

Rock and stone are the most available and ancient of all construction materials. Cut and polished stone is still a preferred material for massive and stately structures in which endurance, solidity, and architectural adaptability are the prime requisites.

Granites are igneous rock solidified from molten magma, although some are derived by regional metamorphism of preexisting rock. Granitization is slow and creates medium- to coarse-grained rock. From St. Nicholas Terrace, a hill on Manhattan island, New York, granite was mined to be the building material for the stately buildings of the City College of the University of the City of New York.

Rapid cooling of magma from volcanoes formed fine-grain rock, such as basalt, diabase, and aplite. These are widely employed in the construction of roads, as prime constituents of concrete, and are known as "trap rock."

Sedimentary rocks were formed from sediments which were themselves products of the disintegration of earlier rocks and soils by water attack, temperature changes, frost, steam and wave erosion, glaciation, landslides, and earthquakes. Millions of tons, even cubic miles, were carried to lakes, rivers, and oceans and were deposited as sediments with calcareous (lime) and siliceous skeletons to form limestone, sandstone, and shale. Rocks have been metamorphosed or changed in form by heat, pressure, and chemical reaction. Granites may be altered to gneisses and schists, sandstone into quartzite, shales into slates (roofing material), and limestone into marble.

The occurrence of building stone is related to the geologic history of an area. The Appalachian district of the eastern United States, stretching from Maine and Vermont to Georgia, is an old mountainous region. The rocks have been gently folded. Metamorphic rocks, marble, slate, granite, and gneisses are found with unaltered granite, sandstone, and limestone.

Between the Appalachians and the Rockies there is a wide area underlain by almost horizontal beds of limestone and sandstone, a reminder that the region was once covered by water. Building limestones predominate in Indiana and building sandstones in Ohio. Granites also occur, notably in Minnesota, Wisconsin and Oklahoma.

In the Rocky Mountain region the rocks are greatly crumpled and folded. On the Pacific Coast side many deposits of igneous, sedimentary, and metamorphosed rocks remind us of vulcanisms of recent geologic age. Lavas, perlite, rhyolite, and volcanic tuffs are common.

Lava is magma or molten rock poured out on the earth's surface at temperatures of 700 to 1200°C.

Perlite is a natural glass with concentric cracks so that the rock breaks into small pear-like bodies, formed by the rapid cooling of viscous lava.

Rhyolites are extrusive igneous (fire) rocks equivalent, volcanically, to granite. Most rhyolites are polymorphic, or "many-formed."

Tuffs are relatively soft porous rocks formed by the compaction and cementation of volcanic ash or dust.

Oliver Bowles[23] had correlated the various building stone dimensions in the United States with the specifications for national, state, and municipal buildings in governmental districts, and according to their origin, occurrence, physical properties, and mechanical preparation for use by man.

Perhaps one of civilization's greatest achievements in the battle to conquer the earth's crust was the development of synthetic rock through the use of cement, concrete, and reinforced concrete, enabling the building of structures such as dams, waterways, bridge foundations, and other massive structures.

Portland cement is an artificial mixture of ground lime-bearing and clay-bearing rocks which is burned to incipient fusion, cooled, and ground to a fine powder. The minerals formed are tricalcium silicate, dicalcium silicate, trical-cium aluminate, and tetracalcium aluminoferrite. Portland cement sets upon hydration by the addition of appropriate quantities of water.

Lehman[24] has discussed cements, including cement types, concrete constituents, aggregates, concrete (massive and reinforced), specifications, properties, usage, and engineering applications. Quantities of cement and concrete far surpass those of any other construction material.

2–8 KARST REGIONS

Karst regions are those areas where carbonate limestone rocks are exposed. Carbonate rocks are sedimentary and must have been uplifted from below sea level high enough to permit fresh water to infiltrate and move through them. The rock lies below or within the local water circulation. In karst regions there is scarcity of soils, as well as of surface streams in rugged topography. There may be an absence of plant life. Once carbonate rocks have been stripped of soil, they remain denuded. Caverns form readily and enlarge, particularly when carbon dioxide dissolves in the water. Limestone caves—sink holes—may become common while the water table lowers itself below the surface. Le Grand[25] has described the problems of karst regions in Jamaica, Puerto Rico, Yugoslavia, Greece, the central United States, the

[23]C. L. MANTELL, *Engineering Materials Handbook*, McGraw-Hill, New York, 1958, Sec. 23, pp. 2–18.

[24]F. G. LEHMAN, in *Engineering Materials Handbook*, ed. C. L. Mantell, McGraw-Hill, New York, 1958, Sec. 25, pp. 1–25.

[25]H. E. LE GRAND, *Science, 179*(4076): (1973).

Nullarbor Plain of Southern Australia, Central Florida, the northern Yucatan Peninsula in Mexico, and the Eastern Mediterranean regions.

It is often assumed that there was a degree of ecological balance in all regions of the world before man changed the landscape with his lumbering, agriculture, dams, engineering structures, and pipeline transmission systems. Le Grand says:

> In karst regions, more than in many other specialized environments, ecological conditions were already skewed and the biota were developed in special and some-times erratic ways. The scarcity of soils, the scarcity of water at the land surface, and the rugged terrain are not conducive to a flourishing and expansive environment.[26]

Some of the problems relating to climate seem almost insurmountable. For example, the immense karst desert of Nullarbor Plain is simply waterless and uninhabitable. To expand significantly the development of organisms seems impossible.

Climate exerts a universally dominant influence on ecology, but processes of karstification have an equally high ecological influence in carbonate rock regions. The development of karst features depends greatly on the degree to which water that contains carbon dioxide has been able to move on and through carbonate rocks and to remove some of the rock as solute. The distinctive features of many karst terrains include scarcity of soils, scarcity of surface streams, and rugged topography; less distinctive are the highly permeable and cavernous rocks, especially at the shallow depths. This high permeability gives rise to many practical problems, including (1) scarcity and poor predictability of groundwater supplies, (2) scarcity of surface streams, (3) instability of the ground, (4) leakage of surface reservoirs, and (5) an unreliable waste-disposal environment.

Natural karst processes in some carbonate rock regions have caused a greater restriction in the development of biota than man can ever be suspected of causing.

[26]Ibid.

THE HYDROSPHERE

3–1 THE WATER CYCLE

Water from the surface of the ocean is evaporated into the atmosphere. The only energy that drives the evaporation is solar radiation from the sun. The solar constant is two calories (1 calorie is the quantity of heat required to raise 1 gram of water 1 degree centigrade) per square centimeter per minute. Over the whole earth the actual amount of radiation is 2.55×10^{18} calories, or about one-fourth of the product of the solar constant and the area of the earth's surface. The ultraviolet rays of the sun's light are absorbed by the ozone of the upper atmosphere; radiation is also reflected by clouds or is scattered at lower altitudes. About 81% of the radiation reaches the earth's surface, but about one-third is reflected back into space. Of the absorbed radiation, 77% is reradiated as long-wave energy back into space. What remains is 23%, which goes into the evaporation of water (Figure 8).

The unending circulation of the earth's moisture and water is the water cycle. It operates in and on the land and oceans as well as in the atmosphere. Water from the oceans, which contain an estimated 97.3% of the earth's water, is evaporated into the atmosphere to form clouds that later condense and fall back to the surface as rain, hail, snow, or sleet. Some of the precipitation, after wetting the ground, runs off the surface to streams. Rapid runoff causes erosion and contributes to floods. Of the water soaking into the ground, some is available to growing plants, some evaporates, and some reaches deeper zones, that is, water layers (termed aquifers, seeps, springs, and groundwater). The streams feed into river systems and eventually flow back to the ocean, as the water runs downhill to sea level.

About 80,000 cubic miles of water per year are evaporated from the oceans and about 15,000 cubic miles from the lakes and land surfaces. Total evaporation equals precipitation, but the condensed matter falls back on the land and the oceans, with 24,000 cubic miles, about one-quarter of the total,

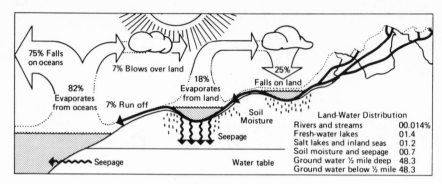

FIGURE 8. The water cycle.

falling on land. This is equivalent to a depth of nearly 500 ft of water over all of Texas.

The distribution of water over the surface of the earth is shown in Table 9. The oceans have better than 97%. Most of the remainder is locked up in glaciers, which have increased, decreased, and fluctuated over geologic ice ages. Underground storage, or aquifers, account for less than 1%, lakes and rivers for a little more than one-tenth of one percent, and the atmosphere for a little more than one-hundredth of one percent. The biosphere, or life and life-endowed organisms, constitute six ten-thousandths of one percent of the water. Water is absolutely essential to life, but has no dependence on human and animal life. We, however, are absolutely dependent on water. We cannot exist on seawater, save that from which we remove the mineral salts. We must have the relatively clean, pure water of underground aquifers, lakes, rivers, and the atmosphere. We adapt, and in isolated areas collect rainwater on the roofs of our habitations and store the water in reservoirs (as in Bermuda). However immense the water cycle might be, we have available for all our purposes less than one percent of the earth's water.

Peixoto and Kettani[1] discussed the water cycle and the effect of the flow of winds in the atmosphere, on a global scale. On the average, whenever there is an imbalance between precipitation and evaporation, there is a need for a net transport of water vapor to or from the locality by atmospheric circulations. An independent check on the precipitation–evaporation difference is available.

For this purpose, time averages of the wind and the humidity will not do, because the effect of the deviations of these factors from their arithmetic mean cannot be neglected. The process must be evaluated instantaneously. An average of the results must be taken over a long interval and for several levels in the atmosphere, up to an elevation of about 5 or 6 km.

[1] J. P. PEIXOTO and M. A. KETTANI, *Scientific American*, 228:4 (1973).

Table 9. Distribution of Water on the Earth's Surface

Region	Water distribution ($\times 10^{15} m^3$)	Percentage
Oceans	1350.	97.3
Glaciers or polar ice	29	1.8
Underground aquifers	8.4	0.6
Lakes and rivers	2	0.15
Atmosphere	0.013	
Biosphere life	0.0006	
Soils	0.066	

The annual mean water-vapor content, averaged for the entire atmosphere, is rather small, amounting to about 0.3% of the total mass of the atmosphere. That quantity is of the same order of magnitude as the amount of water contained in freshwater lakes. Hence the atmosphere is a comparatively small reservoir of water; at a given instant it holds only a tiny part of all the water that participates in the hydrologic cycle. If all the atmospheric water vapor were to condense at once, only a moderate amount of precipitation would result. Even though the total amount of water in the atmosphere is small, however, there is a huge transport of water vapor by atmospheric circulations of various scales in space and in time.

The influence of this small amount of water vapor on the climate of the earth and on hydrologic resources is far out of proportion to its mass. Water vapor plays a major role in the overall energetics of the earth and in the general circulation of the atmosphere. It is the most important factor in all radiative processes of the atmosphere in that it regulates the energy balance through the absorption and transmission of radiation.

Global transport of water vapor by the atmospheric circulation was calculated on the basis of data gathered at numerous stations from pole to pole during the International Geophysical Year (1958). The data consisted of daily measurements of the wind and the water content of the atmosphere obtained with the aid of the IGY aerological radio-sensing network. The averaged results, which show the mean flow of moisture during the year as a field of vectors, represent the lower half of the atmosphere; the upper atmosphere contains only a negligible amount of moisture. The asymmetrical pattern of the vector arrows reflects important irregularities in the global transport of atmospheric water (see Figures 9 and 10).

Mathematical operations were performed on the field of vectors to compare the aerological data with conventional climatological estimates. The outcome is a contour map (Figure 10), in which regions where there is a divergence in the field of vectors (dashed contours on shaded background) correspond to net sources of moisture and hence to places where there is on the

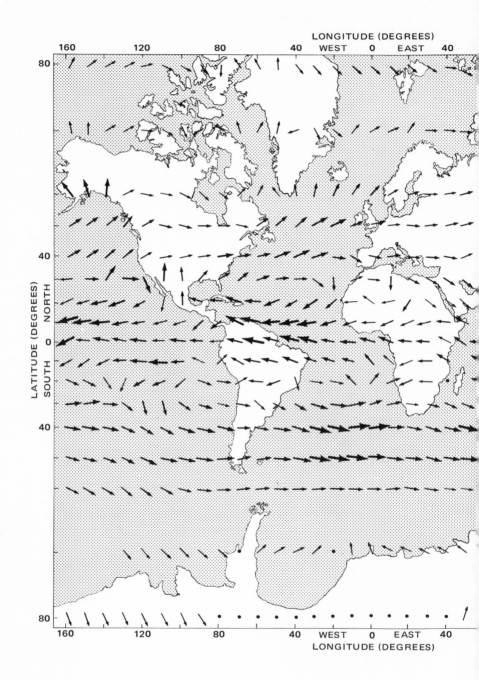

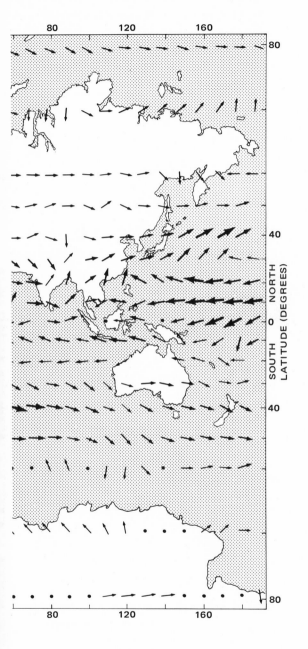

FIGURE 9. Movement of water vapor (hundreds of grams per centimeter per second).

- • LESS THAN 1
- → 1 to 4
- → 5 to 8
- → 9 to 12
- → 13 to 16
- → 17 to 20
- → 21 to 24

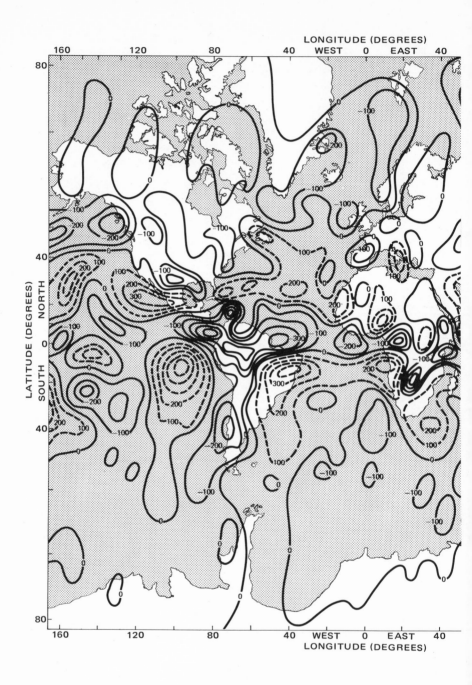

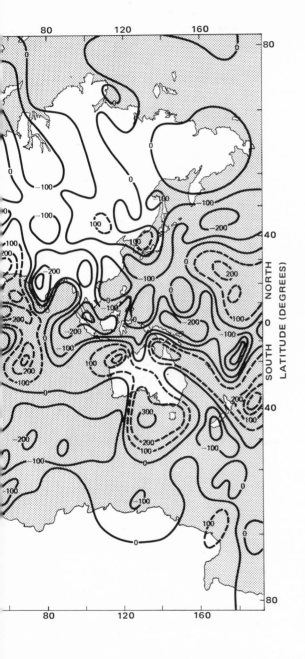

FIGURE 10. Contour map comparing aerologic data with climatological estimates. Dashed contours correspond to where evaporation exceeds precipitation, solid contours to where precipitation exceeds evaporation.

average an excess of evaporation over precipitation, whereas regions where convergence prevails (solid contours on white background) correspond to net sinks of water vapor and hence to places where precipitation exceeds evaporation. The numbers on the contours indicate the equivalent depth of liquid water in centimeters per year.

3-2 THE GREAT LAKES OF THE UNITED STATES

Over geologic time the distribution of terrestial water has changed. When continental ice caps formed at maximum glaciation, the sea level may have been lowered by as much as 140 m, exposing large areas of the continental shelves. If the ice caps were to melt, the sea level would rise as much as 60 m, flooding a large area of the continents. Against such changes man must retreat to higher ground if it is available.

Ages ago, an extensive ice sheet covered the area that is now Ontario, Canada, and Michigan, Minnesota, Wisconsin, and adjoining states. In its advance and later retreat the Great Lakes were formed, presenting a watery staircase that spans 800 miles. Lake Superior, the highest and deepest lake, lies 21 ft above Lakes Michigan and Huron. A drop of 7 ft separates these lakes from Lake Erie, which in turn is 325 ft higher than Lake Ontario. The surface areas, depths, and volumes of the Great Lakes are given in Table 10.

At times the lakes had other outlets. Water drained through the Mississippi River, as well as via the Finger Lakes in New York and the Susquehanna River, or via the Mohawk Valley and the Hudson. It is estimated that the St. Lawrence River became the major path about 4000–12,000 years ago.

The Great Lakes is an area where sailors have been dominant in humanizing the watery region. From Lake Superior, boats bypass rapids of the St.

Table 10. The Great Lakes[a]

Lake	Surface area (square miles)	Volume (cubic miles)	Average depth (feet)	Connecting river or strait	Rank over earth	Greatest depth (feet)
Superior	31,700	2,938	489	St. Mary's	2	1,333
Michigan	22,300	1,180	279	Mackinac	6	923
Huron	23,000	849	195	St. Clair and Detroit	5	750
Erie	9,910	116	62	Niagara	12	210
Ontario	7,340	393	283	St. Lawrence	14	802
Total	94,250					

[a]For comparison: Caspian Sea surface area is 143,550 square miles.

Mary's River at the locks of the Sault Ste. Marie canals. The boats pass from Lake Huron to Lake Erie via the St. Clair River, Lake St. Clair, and the Detroit River. Connection with Lake Michigan is by the Straits of Mackinac. Passage from Lake Erie to Lake Ontario is via the Welland Canal, whose eight locks conquered the Niagara escarpment. From Lake Erie, passage is via the Thousand Islands in the St. Lawrence, the Eisenhower Locks at Massena, and the St. Lawrence Seaway to the ocean. Thus the Great Lakes cities became seaports through our engineering skill.

More than 25% of United States income from farming, mining, and manufacturing is developed in the Great Lakes area. The farmers were able to use their fertile fields only after draining. The U.S. Geological Survey and the Departments of Agriculture and the Interior have often strongly rebutted the statement of some environmentalists that "Lake Erie is dead." Lake Erie is the shallowest of the Great Lakes and was created by retreating glaciers some 20,000 years ago. Barring another ice age, Erie has several thousand years to go before senility. The western part of the lake receives a large amount of natural organic material transported from the surrounding terrain, and that is where the algal growth has always been present, in the extremely shallow areas. Over the last 100 years, Lake Erie has consistently produced about 50% of the fish catch of the entire Great Lakes system, according to Department of Census and manufacturers' figures. This is not the mark of a dead lake.

Regier and Hartman[2] confirm the statement that Lake Erie's biotic community is not dead. Their tables for all varieties and species of fish show that catches have increased, although there may be fewer of the species most popular in the 1950s. They also provide an extensive bibliography. Intergovernmental cooperation and monitoring methods exist for complete restoration of Lake Erie's condition to the 1840–1850 status.

Lakes are depressions in the crust of the planet which collect water from atmospheric precipitation such as rain, melting ice and snow, springs, rivers, and runoff from land surfaces. They are widely distributed, but more abundant at high altitudes and in recently glaciated mountain regions. They are frequent along rivers of low gradient and wide flats. They occur at all altitudes, as shown in Table 11.

Formation of lakes may be the result of changes in the earth's crust by physical forces, as shown in Table 12.

From the time of its birth or formation, a lake is destined to die and disappear. Even when there is enough water to keep the lake filled, there is deposition of detrital matter which fills the lake, and the effluent stream cuts deeper and deeper channels which in turn lower the lake.

[2]H. A. REGIER and W. L. HARTMAN, Lake Erie's fish community—150 years of cultural stress, *Science, 180* (4092): 1248–1255 (1973).

Table 11. Range of Lake Elevations

Lake	Elevation (ft above sea level)	Elevation (ft below sea level)
Titicaca, Bolivia	12,500	—
Yellowstone, Nevada, USA	7,741	—
Superior, USA	500	—
Caspian Sea, USSR	—	86
Sea of Tiberias, Israel	—	682
Dead Sea, Israel	—	1,392

Table 12. Geologic Origin of Lakes

Physical force	Example	Lakes
Glacial erosion and terminal basins	Lower valleys of the Alps, Switzerland	Neuchatel, Lucerne, Zurich
Valleys deepened by glacial erosion	Michigan, Great Lakes, Wisconsin	Superior, Huron, Michigan
Landslides causing barriers	California Lakes	—
Damming by valley glacier	Aletsch Glacier, Switzerland Malaspina Glacier, Alaska	Mäyllen Costain
Blocking by frontal moraines	Canadian Lakes	St. John, Quebec
Alluvium build-up	Lakes formed by levees of Mississippi	—
Volcanic craters	Ancient and dormant volcanoes	Ngo, Japan
Synclines	Joux Jura	Switzerland
Rift valleys	Great Rift Valley, North Africa	Tanganika, Nyassa
Sinkholes	Limestone caves and clay regions	—
Rise of water table	Desert oases	—

When rainfall is less than 10 inches per year, evaporation losses are greater than precipitation gain, and salt and bitter lakes result. The Dead Sea and the Great Salt Lake in Utah had fresh water ancestors, while the Caspian and Aral Seas were once part of a much larger portion of the ocean. Uplift of the Carpathian mountains greatly reduced their area.

Lakes may become stagnant. Influx of organic matter, anaerobic decomposition, and excess of nutrients with stimulation of algal growth and death, have been studied by biologists and limnologists. Biologic oxidation by pumping air or pure oxygen eliminates stagnant areas, and restores circulation and environmentally healthful conditions. This treatment is being applied and studied in the Lakes area of New York at Lake Waccabuc. The Great Lakes of North America are among the largest over the world's entire crust, as seen from Table 13. Other large United States lakes are listed in Table 14.

Table 13. Large Lakes of the World
(*Source: National Geographic Society*)

Lake	Continent	Area (square miles)	Length (miles)	Maximum Depth (feet)
Caspian Sea	Asia–Europe	143,550	760	3,264
Superior	North America	31,800	350	1,333
Victoria	Africa	26,828	750	265
Aral Sea	Asia	25,800	280	223
Huron	North America	23,000	206	750
Michigan	North America	22,400	307	923
Tanganyika	Africa	12,700	420	4,710
Great Bear	North America	12,275	192	1,356
Baikal	Asia	11,750	395	5,315
Nyasa	Africa	11,430	360	2,226
Great Slave	North America	10,980	298	2,015

Table 14. U.S. Natural Lakes other than the Great Lakes
(*Courtesy USGS*)

Lake	Location	Area (square miles)
Lake of the Woods	Minnesota, Ontario, and Manitoba	1,695
Great Salt	Utah	1,000
Iliamna	Alaska	1,000
Okeechobee	Florida	700
Pontchartrain	Louisiana	625
Champlain	New York, Vermont, and Quebec	490
St. Clair	Michigan and Ontario	460
Becharof	Alaska	458
Red Lake	Minnesota	451
Salton Sea	California	350

Table 14a. Largest Man-Made U.S. Lakes
(*Courtesy, USGS*)

Lake	Location	Surface Area (square miles)	Maximum Depth (feet)
Sakakawea	North Dakota	609	200
Kentucky	Kentucky and Tennessee	408	145
Fort Peck	Montana	383	220
Powell	Arizona and Utah	252	580
Mead	Arizona and Nevada	246	589

Lakes, brooks, springs, and aquifers feed from higher elevations to rivers, and from rivers to the sea. Rivers have flood plains over which some sort of human control has been exerted ever since recorded history began. Dams created artificial lakes; the largest ones in the United States are listed in Table 14a. The world's highest dams are listed in Table 15, the largest dams in Table 16, the world's greatest man-made lakes for flood control and associated hydroelectric power, and the world's largest hydroelectric plants, in Tables 17 and 17a, respectively.

Table 15. The World's Highest Dams

Name	Location	Height (feet)	Earth	Concrete	Completed
Nurek	USSR	1,040	X		
Inguri	USSR	860		X	
Grand Dixence	Switzerland	932		X	1962
Vaiont	Italy	858		X	1961
Mica	Canada	800	X		
Mauvoisin	Switzerland	777		X	1958
Sayansk	USSR	774			
Oroville	USA	770	X		1968
Chirkey	USSR	764		X	
Contra	Switzerland	754		X	1965
Bhakra	India	740		X	1963
Hoover	USA	726		X	1936

Table 16. The World's Largest Dams
(*Courtesy, USDI*)

Name	Country	Volume (million cubic yards)	Completed
Tarbela	Pakistan	159	
Fort Peck	USA	126	1940
Mangia	Pakistan	108	1967
Oahe	USA	92	1963
South Saskatchewan	Canada	86	1966
Oroville	USA	80	1968
San Luis	USA	78	1967
Nurek	USSR	76	
Nagajunasagar	India	74	1966
Garrison	USA	67	1956

Table 17. The World's Greatest Man-Made Lakes

Name	Capacity (million acre-feet)	Country	Completed
Owen Falls	166	Uganda	1954
Bratsk	137	USSR	1964
Kariba	130	Rhodesia–Zambia	1959
Sadd-El Aali	126	UAR	
Akosombo	120	Ghana	1965
Manicouagan #5	115	Canada	1967

Table 17a. The World's Largest Hydroelectric Plants

Name	Country	Initial Operation	Ultimate Installed Capacity (mw)
Sayansk	USSR		6,300
Krasnoyarsk	USSR	1967	6,000
Grand Coulee	USA	1941	5,574
Bratsk	USSR	1961	4,500
Sukhovo	USSR		4,500
Churchill Falls	Canada		4,500

3–3 RIVERS AND FLOOD CONTROL

In the flood control of rivers, we have won battles with great flows of water over tremendous distances. The major United States rivers are listed in Table 18, prepared by the Army Corps of Engineers, which controls navigation on them. Table 19 lists the major foreign rivers, using data from the National Geographic Society.

Rivers play a significant role both in the Earth's hydrologic (water) cycle and in its petrogenic (rock) cycle, emptying water from the land into the oceans and gradually wearing away the land in their drainage systems by carrying sediment to the sea. Moreover, rivers have historically offered the most convenient pathways for exploration, later becoming courses for transportation and commerce. Civilizations based on agriculture have always flourished along fertile river valleys. The flood and flow of many river systems are now controlled by dams and reservoirs to assure most efficient use of the water.

The longest rivers of the world, in miles, are: the Nile (4145), Africa; the Amazon (3900), South America; the Mississippi–Missouri–Red Rock (3710), United States; the Ob–Irtysh (1460), Siberia; the Yangtze (3400), China; the

Table 18. Major United States Rivers
(Courtesy, Army Corps of Engineers)

River	Empties into	Length (miles)
Mississippi–Missouri–Red Rock	Gulf of Mexico	3,741
Rio Grande	Gulf of Mexico	1,885
Yukon	Bering Sea	1,875
Arkansas	Mississippi river	1,450
Colorado	Gulf of California	1,450
Ohio–Allegheny	Mississippi river	1,306
Red (Oklahoma–Texas–Louisiana)	Mississippi river	1,222
Columbia	Pacific Ocean	1,214
Brazos	Gulf of Mexico	1,210
Snake	Columbia river	1,000

Table 19. Major Foreign Rivers
(Source: National Geographic Society)

River	Location	Empties into	Length (miles)
Nile	Africa	Mediterranean	4,145
Amazon	Brazil	Atlantic Ocean	3,900
Ob–Irtysh	USSR	Gulf of Ob	3,460
Yangtze	China	East China Sea	3,400
Yellow	China	Yellow Sea	3,00
Congo	Africa	Atlantic Ocean	2,718
Amur	USSR	Tatar Strait	2,700
Lena	USSR	Lapter Sea	2,680
Mackenzie–Peace	Canada	Beaufort Sea	2,635
Mekong	S.E. Asia	South China Sea	2,600
Niger	Africa	Gulf of Guinea	2,600

Huang or Yellow River (3000), China; the Congo (2718), Africa; the Amur (2700), Siberia; the Lena (2680), Siberia; the Mackenzie–Peace (2635), Canada; the Mekong (2600), Southeast Asia; the Niger (2600), Africa; the Parana (2500), South America; The Murray–Darling (2310), Australia; and the Volga, (2290), East Europe. The Angara is the only outlet of Lake Baika USSR, and drops 1140 ft in 1150 miles.

Most fish spawn and live their first life in the shallower parts of the sea adjacent to the coast. Fish migrate into fresh water to spawn and lay eggs in the brackish waters of the intertidal estuarine zones, the marshlands, and the tidal estuaries adjacent to the sea. Ocean fish of many species actually return to the shallow bays to spawn.

In the eastern Mediterranean a dramatic change has occurred since the completion of the high Aswan Dam in Egypt, in 1965. Previously, the annual discharge of fresh floods into the Mediterranean from the Nile was about 80,000 million cubic meters per year. This flow came chiefly in late summer and early autumn as a reflection of the monsoon rains at the headwaters of the Blue Nile in Ethiopia, 2500 miles upstream. The dam, which is some 364 ft high, has formed Lake Nasser, with a total area of 1.25 million acres. It holds back all but about 5% of the seasonal discharge of Nile flood waters.

The Nile River floods once carried huge quantities of silt which were a source of nutrients and minerals and contained chemicals washing out of the soil itself and the agricultural and sewage products of more than 70 million people. The Nile Valley was thus provided with enrichment which reached the sea each year after flooding the adjacent lowlands along the delta. After the flood there was a noticeable outburst of "bloom" on the Mediterranean—an immense reproduction of algae and other plant and animal organisms. A similar bloom in Lake Erie is called "pollution" by environmentalists.

This rich bloom, with some 50–100 million tons of silt that flooded into the brackish estuarine waters of the delta and the adjacent shallow sea, was a most effective marine fertilizer. It nourished more than one-half the fish caught annually by Egyptians. The end products of this food chain were Tilapia fish in the rivers, and mullet and other edible saltwater species, as well as sardines, further offshore. The sardine take alone was approximately 18,000 tons per year.

The blooms of algae caused by the annual enrichment have virtually disappeared. To the environmentalist, the Aswan Dam has removed pollution from the Mediterranean. Sardines have become scarce, and the catches have been reduced to some 500 tons per year, less than 3% of the former crop. The delta itself is being eroded. Some of the delta lakes have become less saline, lacking the blend of fresh and salt waters for ideal brackish conditions. The fish and shrimp that were formerly abundant in these shallow estuaries are unable to spawn.

Rivers in the United States are often called dirty and polluted. Rivers are the natural transport system for sediment, organic matter, and drainage of the rains and melting snow that fall on the land. Other than in the Middle Atlantic and New England states, and the Pacific Northwest, rivers are not clean, limpid streams, but colored and muddy. "Big Muddy," the Missouri, carries a load of more than 2 million tons of sediment per day into the Gulf of Mexico, or over 700 million tons per year. The daily load of sediment is the equivalent of 40,000 freight cars. The Savannah is brown and muddy, as are the Pee Dee, the Warrior, the Ohio, the Colorado, the Atchafalaya, and the Rio Grande. These augment the load of detritus carried to the oceans. The exceptions are rivers like the St. Lawrence, with its tributaries, the Niagara, the Shawinigan, the Sagenay, and the Columbia and the Hood river systems. These are clear, sparkling, without mud or color.

Table 20. The World's Greatest Waterfalls
(*Courtesy, United States Geologic Survey*)

Continent	Country	Location	Height (ft)
Africa	Ethiopia	Fincha	508
	Lesotho	Maletsunyane	630
	South Africa	Aughrabies	400
		Tugela	3,110
	Tanzania–Zambia	Kalambo	726
	Zambia	Chirombo	800
Asia	India	Cauvery	330
		Gersoppa	830
	Japan	Kegon	330
		Yudaki	335
Australia	New South Wales	Wentworth	518
	Queensland	Wollomombi	1,100
New Zealand		Sutherland	1,904
Europe	Austria	Krimoni	1,250
	France	Gavarnie	1,385
	Norway	Mardalsfoss (Eastern)	1,696
		Veltis	1,214
	Switzerland	Cascade de Giétroz	1,640
		Giessbach	1,312
		Trümmelbach	1,312

Rivers in their passage to the sea create some lakes where the drop in level is small, but create large waterfalls where the change in elevation above sea level is precipitous. There are tens of thousands of waterfalls throughout the globe. Some have been humanized to furnish power, some are in national parks, and some are "wild rivers" in undeveloped areas or in scarcely inhabited wildernesses. The highest waterfalls in the world are: Angel (Venezuela): 3212 ft; Tugela (South Africa): 3110 ft; Yosemite (California): 2425 ft; Cuquenan (Venezuela): 2000 ft; Sutherland (New Zealand): 1904 ft; Eastern Mardalsfoss (Norway): 1696 ft (Table 20).

On the basis of annual flow combined with considerable height, Guaira, between Brazil and Paraguay, is the world's greatest waterfall; its estimated annual flow is 470,000 cubic feet per second (cusecs). A greater volume of water passes over Stanley Falls, in the Democratic Republic of the Congo, but not one of its seven falls, spread over a distance of 60 miles, is over 10 ft. The estimated annual flows of other great waterfalls, in cusecs, are: Niagara (Canada and the United States), 212,200; Paulo Afonso (Brazil), 100,000; Urubupunga (Brazil), 97,000; Iguazu (Argentina and Brazil), 61,600; Patos Maribondo (Brazil), 53,000; Victoria (Zambia and Rhodesia), 38,400; Churchill (Labrador), 30,000–40,000; and Kaieteur (Guyana), 23,400.

Table 20 *(continued)*

Continent	Country	State or Province	Location	Height (ft)
North America	Canada	British Columbia	Takakaw	1,650
			Panther	600
		Labrador	Churchill	245
		Mackenzie District	Virginia	315
		Quebec	Montmorency	251
	Canada–USA	Ontario	Horseshoe	186
		New York	American	193
	United States	Arizona	Mooney	220
		California	Feather	640
			Illilouette	370
			Nevada	594
			Ribbon	1,612
			Silver Strand	1,170
			Vernal	317
			Yosemite	2,425
		Colorado	Seven	266
		Georgia	Tallulah	251
		Idaho	Shoshone	195
		New York	Taughannock	215
		Oregon	Multnomah	620
		Tennessee	Fall Creek	256
		Washington	Fairy Falls	700
			Palouse	198
			Sluiskin	300
		Wisconsin	Manitou	165
		Wyoming	Yellowstone (Upper)	109
			Yellowstone (Lower)	308
	Mexico		El Salto	218
South America	Brazil		Glass	1,325
	Guyana		King George VI	1,600
	Colombia		Catarata de Candelas	984
	Venezuela		Angel	3,212
			Cuquenan	2,000

3-4 THE TENNESSEE VALLEY AUTHORITY AND FLOODS

The Tennessee Valley Authority (TVA) was created over 40 years ago, in the summer of 1943, by a generation of conservationists, Congress, and President Franklin Roosevelt, primarily as a flood-control agency and only secondarily as a power producer. The creation of a huge system of dams, lakes, and flood-control projects was successful. It is shown in Figure 11.

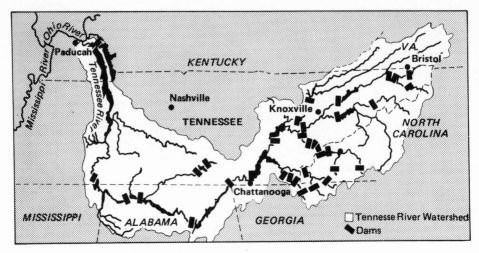

FIGURE 11. The Tennessee Valley region, showing watershed and dams.

At first, TVA was a hydraulic power producer, with low rates to attract industry. Customers grew faster than the hydraulic power supply, and TVA moved into coal-fired plants to satisfy demand. Coal power, not water power, now accounts for 80% of the 100 billion kwh per year produced by TVA, and feeds to 160 local electric systems and ultimately to 2.3 million customers. In 40 years TVA has gone from a successful and unified river control system to an establishment of 21.9 MW of installed generating capacity, making it among the largest, if not the largest, in North America.

TVA 1970 estimates show the annual bill for a typical Tennessee all-electric home at $292 for 24,000 kwh, or $25 per month, equivalent to about one dollar per day for heating, cooking, and household appliances. TVA maintains that it took coal furnaces out of every house in Knoxville. When TVA began operations, emissions and dust were common. TVA is now the country's largest coal consumer of mountain strip mining coal, burning 35 million tons a year.

Southern California Edison figures that snow runoff to its hydroelectric plants in the southern Sierra Madre, helped by rainfall produced by cloud-seeding, has increased 8% since 1951.

When the water cycle deviates severely from its historical custom, people and human works may be destroyed by floods from rivers too heavily loaded by sudden or violent rain, or by the results of melting snow from record-breaking snowstorms. Ten inches of snow is equivalent to 1 inch of rain. Dams are designed only to accommodate the maximum flows of a 40-year cycle. Floods can be more destructive than volcanoes, but the damage they cause is generally small in comparison with earthquakes, although tidal waves, which are themselves associated with submarine earthquakes and/or volcanoes, have killed hundreds of thousands of people. Tables 21, 22, and 23 provide some history.

In the United States, we have controlled floods in such large areas as the Tennessee River basin, but, in 1973, the Mississippi flooded vast areas because flood-control dams, levees, locks, and bank stabilizations still do not exist on a sufficiently grand scale. Coupled with Hurricane Camille, the Susquehanna in Pennsylvania ripped out bridges and flooded many low areas. Wilkes-Barre was still stricken more than a year later; curiously, rebuilding continues on the flood plain of the river!

In January, 1969, Southern California had torrential rains which pounded the Los Angeles–Santa Barbara area. The area suffered the most damaging flood in 31 years, with rain for more than a week. When the rains finally ended, property damage was more than $60 million, 9000 homes were destroyed, and 91 persons had died.

The Mississippi is the trunk of a river system whose branches are the Missouri, the Red, the Ohio, the Tennessee, the Arkansas, the Des Moines, the Cedar, the Rock, the Illinois, the Wisconsin, the Wabash, the Platte, and many other rivers. These draw the runoff rains, the melting snow, and drainage from

Table 21. World's Principal Floods

Date	Location	Deaths	Explanation
1228	Netherlands	100,000	Seaflood to Friesland
1642	China	300,000	Kaifeng seawall destroyed
1787	Eastern India	10,000	Storm drove seawater inland 20 mi
1887	Honan, China	500,000	Yellow River overflowed
1889	Johnstown, Pa.	2,000	Flood; dam breakage
1900	Galveston, Texas	6,000	Tidal wave
1911	China	100,000	Yangtze River overflowed
1939	China	1,000,000	Floods in north; extensive drownings and starvation
1947	Honshu, Japan	2,000	Flooding after typhoon
1948	Foochow, China	1,000	Floods
1951	Manchuria	1,800	Flooding
1953	Northwest Europe	1,794	Storms and floods devastated North Sea coastal areas
1954	Iran	2,000	Flash flood
1955	Pakistan and India	1,700	Flood
1956	China	2,000	3 Provinces flooded after typhoon
1957	Western Mexico	2,000	Flood
1960	East Pakistan	10,000	Tidal wave
1963	Northern Italy	2,000	Vaimont Dam collapsed
1968	Western India	1,000	Widespread flooding; killed 80,000 cattle

Table 22. Heaviest Snowstorms and Snowfall in the United States

(*Courtesy, U.S. Dept. of Commerce*)

Date	Location	Depth (inches)	Description
2/19–24/1717	New England	60–72	Drifts covered many one-story houses
12/3–6/1886	Asheville, N. Car.	26	Heavy snow and high winds
3/11–14/1888	Boston, New York City, Philadelphia, and Washington	40–50	Major eastern cities paralyzed
2/14–15/1894	Rayne, Louisiana	24	Record for Louisiana
2/11–14/1899	Washington, D.C.	34.2	Record for Capital
4/19–21/1901	Watertown, Ohio	45	Record 24-hour fall for state
11/7–11/1913	Ohio, Pennsylvania, and West Virginia	36	Severe snow and windstorm
12/29–31/1915	Flagstaff, Ariz.	54	Record for state
12/23–26/1916	Summit, Calif.	59	Heaviest 24-hour fall in California
4/14–15/1921	Silver Lake, Colo.	95	75.8 inches in 24 hours; greatest in U.S. history
11/17–20/1921	The Dalles, Oregon	54	Record in state
4/27–28/1928	Bayard, West Virginia	35	Record 24-hour of 34 inches
1/18–19/1933	Giant Forest, California	60	Record for state for 24 hours
1/18–22/1940	Watertown, New York	69	Record for state
11/2–6/1946	New Mexico	36	Record for state for 24 hours
3/2–5/1947	Readsboro, Vermont	50	Record for state
3/2–5/1947	Peru, Massachusetts	47	Record for state
1/1–6/1949	Colorado, Idaho, Montana, Nebraska, North Dakota, South Dakota, Utah, and Wyoming	36–48	Most adverse weather in history of the West
3/25–27/1950	Dumont, South Dakota	60	Record for state
11/23–28/1950	Ohio, Pennsylvania, and West Virginia	33–57	Record for 3 states
1/11–16/1952	Tahoe, California	149	Greatest for single storm in U.S. (except Alaska)
3/2–5/1955	Colorado, Idaho, Montana, South Dakota, and Wyoming	30–52	One of the heaviest and latest spring storms
2/18–20/1960	Maryland, New England, New Jersey, New York, Pennsylvania, and West Virginia	20–36	Damage in millions of dollars
2/2–5/1964	New Mexico, Oklahoma, and Texas Panhandle	18–36	Second heaviest storm in Panhandle's history
2/24–27/1969	New England	25–60	Most severe storm in many areas for years
2/10–11/1973	Georgia, North Carolina, and South Carolina	15–21	Worst storm in South in century

Table 23. World's Greatest Rainfall
(*Courtesy, U.S. Dept. of Commerce*)

Date	Location	Duration	Depth (inches)
7–8/—/1860–61	Cherrajuni, India	1 year	1,041.78
7/1/1861	Cherrajuni, India	1 month	366.14
5/12/1916	Plumb Point, Jamaica	15 minutes	7.80
6/22/1947	Holt, Missouri	42 minutes	12.00
3/15–16/1952	Cilaos, Reunion	24 hours	73.62
7/4/1956	Unionville, Maryland	1 minute	1.23
2/28–29/1964	Belouve, Reunion	12 hours	42.76

31 states and Canada, in an area from the Rocky Mountains and the Great Divide to the Appalachians in the East.

Heavy snows in the winter of 1972, prolonged rains in the Midwest, and freak spring blizzards in the upper river states caused, through April and May, 1973, the entire system from Minnesota to New Orleans to crest above flood stage. For a brief period, the waters receded slightly, and then, fed by more rain and snow, crested higher. This occurred four times. Millions of acres of farmland were covered with water. Dozens of river towns, summer resorts, fishing camps, industrial plants, urban housing areas, railroads, and county, state, and interstate highways were submerged. Ferries were put out of action, bridges demolished, and the central granary of the nation paralyzed. Levees, the earthwork barriers with their hold-in-place revetments, built up over the years to hold back the currents of the river, broke or washed out. Those that held had water cascading over them. The 27 flood-control dams above St. Louis and their navigation locks were like flotsam as seas of water spread 3–4 miles beyond their ends. The suburbanites, who had been assured by developers that the river would never dampen their doors, found that their ruined homes (those which did not float away) were situated in the flood plain of a mighty river. The longest and oldest dam, at Keokuk, built primarily for power generation and to make the rapids navigable, had to open its spillways to relieve crushing pressure upstream from ice jams. When the ice and snow came down in late April, the industries below the dam were flooded. Quincy, Illinois, a few miles downstream, was a disaster area. At Hardin, Illinois, the bridgemaster opened a drawbridge to let a house go through.

Casualties were few because the Mississippi system never floods by surprise. Everybody knew it was coming, but meteorologists could not predict with what severity. In time, the water and the river went down. Granaries and elevators held spoiled grain and beans, but planting was begun as soon as fields were drained and again ploughable. The people went back to their fields, homes, factories, and cities to clean them up and make them useful again. The next efforts at flood control of the Mississippi and Missouri will be drawn up on the basis of a 100-year cycle to protect against further such serious disasters.

On June 9, 1972, 14 inches of rain spilled onto a small area of the Black Hills of western South Dakota. Normally, 14 inches fall in an entire year (Figure 12). In the late afternoon a strong breeze blowing from the Southeast carried an unusually moist supply of air to the eastern side of the Black Hills. The steep slopes forced the incoming air upward, causing moisture to accumulate over the hills. Normally, high-level winds would carry much of this moisture away, but on this day the upper-level circulation had come to a near standstill. The damp accumulation hovered, almost motionless, and the rains

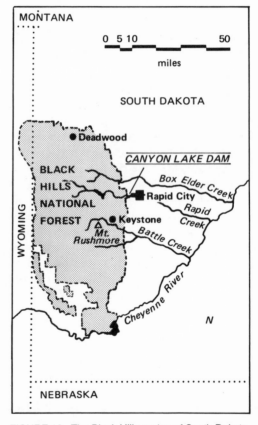

FIGURE 12. The Black Hills region of South Dakota.

began to fall. The steep, rocky, 3500- to 7000-ft-high Black Hills could not absorb the downpour. From near the foot of Mount Rushmore, the runoff picked up battering-ram force as it funneled through narrow canyons and dropped toward the grasslands to the east. Thundering against the back door of Rapid City, where 43,000 people lived, the flash flood crumpled a 34-year-old earthen dam and unleashed a rampaging 5-ft-high wall of water through the heart of the town.

Before the rains stopped and the flood dissipated, 237 people had died, 5 were missing, and 5000 had been left homeless in a 30-mile-long, half-mile-wide path of destruction. Dozens of bridges were destroyed, 5000 cars were demolished or damaged, and 1200 homes and about 100 business buildings had vanished.

3-5 DRAINAGE OF WETLANDS

We often hear the cry "Save our wetlands, or we will damage our environment." Yet the Hollanders for centuries have been stimulated by an exactly opposite philosophy: "Recover our wetlands from the sea." They improved their fisheries and seafood production. Wetland areas are highly variable with changes of season, rain, hail, snow, tides, and ocean currents, so much so that in Virginia and Maryland they have had to be controlled in order to maintain the lobster, clam, crab and fish industries.

Since the early days of settlement, American farmers have drained land.[3] In 1763, the Dismal Swamp area of Virginia and North Carolina was surveyed by George Washington and others with a view to land reclamation and inland water transportation. The Dismal Swamp Canal Company was chartered, in 1787, by the two states. The canal was opened 7 years later. It is still a means of transportation and helps to prevent floods.

Early drainage works were also constructed in Delaware, Maryland, New Jersey, Massachusetts, South Carolina, and Georgia. Drainage work of concern to the public was carried out under the authority of colonial and state laws. Among the earliest laws are those of Maryland, in 1790, and Delaware, in 1816. The North Central states likewise became interested in land drainage shortly after settlement. Michigan and Ohio had drainage laws by 1847.

The early work which started in colonial days consisted primarily of constructing small open ditches to drain wet spots in fields, and of cleaning out small natural streams. Little engineering work was involved. In 1835, John

[3]*Water, The 1955 Yearbook of the Department of Agriculture*, Government Printing Office, Washington, D.C., 1956.

Johnson of Seneca County, New York, brought over, from Scotland, patterns from which clay tile was molded by hand and laid on his farm. This was the beginning of modern drainage in the United States.

Settlement of the Ohio and Mississippi valleys was just starting. Much of this land, though very fertile, could not be cultivated until it had been drained. Malaria was prevalent in large areas. Here the use of tile spread rapidly; 1140 tile factories, mainly in Illinois, Indiana, and Ohio, were in operation by 1880. More than 30,000 miles of tile had been laid in Indiana by 1882.

Farmers learned that the success of many tile systems depended on large outlet ditches. The construction of such ditches increased rapidly as the North Central states were settled. The Ohio Society of Engineers and Surveyors reported, in 1884, that 20,000 miles of public ditches had been constructed in Ohio, benefiting 11 million acres of land and improving the health of the citizens.

Drainage has added an estimated 25–30 million acres to the tillable areas in the North Central states, and has increased production on about 37 million acres more. Typical of the additions to the tillable land are large tracts in northwestern Ohio, northern Indiana, north central Illinois, north central Iowa, and southeastern Missouri. What the tracts were once like is described in the report on Long's expedition to the source of the Minnesota River in 1823, written by W. H. Keating, of the University of Pennsylvania, who accompanied the expedition. The Long expedition, undertaken by order of J. C. Calhoun, Secretary of War, sought information about the conditions in the country, then undeveloped, west of Pennsylvania.

Professor Keating described the land east of Fort Wayne, Indiana, thus:

> Near to this house we passed the State line which divides Ohio from Indiana. . . . The distance from this to Fort Wayne is 24 miles, without a settlement; the country is so wet that we scarcely saw an acre of land upon which settlement could be made. We travelled for a couple of miles with our horses wading through water sometimes to the girth. Having found a small patch of esculent grass (which from its color is known here as bluegrass) we attempted to stop and pasture our horses, but this we found impossible on account of the immense swarms of mosquitoes and horse flies which tormented both horses and riders in a manner that excluded all possibility of rest.

He also described the land south and west of what is now Chicago:

> From Chicago to a place where we forded the Des Plaines River the country presents a low, flat and swampy prairie, very thickly covered with high grass, aquatic plants, and among others, the wild rice. The latter occurs principally in places which are under water; its blades floating on the surface of the fluid like those of the young domestic plant. The whole of this tract is overflowed during the spring and canoes pass in every direction across the prairie.

Drainage changed those conditions. Today, the traveler who notes the well-cared-for, productive fields, the substantial farm buildings, the good roads, and the splendid school buildings, may not think that drainage made possible many of the developments, that without drainage the localities would be much the same as Keating described them in 1823. In both areas there are now more miles of public outlet ditches and drains than there are miles of public highways.

The Swamp Land Acts of 1849 and 1850 were the first important federal legislation relating to land drainage. They were the result of more than 20 years of discussion in the Congress of appropriate procedures for initiating reclamation of the wetlands of the public domain. For more than 75 years they were almost the only statement of federal drainage policy. Under the acts, vast acreages of swamp and overflowed lands were transferred to the states on the condition that funds from their sale be used to build the drains and levees necessary to reclaim them.

Under the Swamp Land Acts of 1849, 1850, and 1860, approximately 64 million acres of swamp and overflow land in 15 states were turned over to the respective states to facilitate reclamation of the land for agricultural use. No important reservations were attached to this transfer, and the states were free to dispose of the land as they saw fit. In that way the Federal Government relinquished control of most of the potential drainage work in the public domain.

It has become common practice to dismiss as failures the drainage and flood-control projects started under the Swamp Land Acts. It is true that for the most part the states did not immediately develop the land as anticipated. But that is not the whole story.

Over the lower Mississippi Valley states, where administration and use of swampland funds was a major political, economic, and social issue for more than 30 years, reclamation carried out under the Swamp Land Acts permanently affected the agricultural economy. The experience in flood control and drainage engineering gained in trying to meet the provisions of the grants formed the basis for the elaborate drainage projects later undertaken by local districts and by the states and the Federal Government for control of floods in the lower Mississippi Valley. Likewise, most of the legal and administrative concepts and machinery set up under the Swamp Lands Acts became a permanent part of flood-control and drainage practices.

Disastrous floods occurred frequently from 1850 through 1900. Loss of life and property led to increased expenditures for flood-control work each year. Flood-control improvements in time grew to an impressive size. With their increasing enlargement, waves of optimism found expression in the development of new land. By 1900, it appeared that public reclamation work should include land drainage as well as levee building. Soon the landscape of much of the Delta region was altered through drainage activities initiated and financed through the creation of local drainage districts under state laws.

Drainage is a land improvement and cultural practice on about 2 million farms. The agricultural census of 1950 reported nearly 103 million acres of land in organized district and county drainage enterprises in 40 states. More than $900 million, or an average of about $9 per acre, has been expended on public drainage improvements on the 103 million acres, which is larger than the combined areas of Ohio, Indiana, and Illinois. More than 155,000 miles of outlet ditches, 56,000 miles of main-outlet tile drains, 7800 miles of levees, and pumping plants of more than 110,000 horsepower have been constructed. The enterprises range in size from less than 100 acres to more than 1 million acres; they average about 7300 acres. Of the 103 million acres in drainage enterprises, some 15 million acres are still too wet for cultivation; crop losses are frequent on an additional 10 million acres because of poor drainage.

The map, Figure 13, and Tables 24 and 25 show where the drained lands are located and the rate at which drainage is done. Land in organized drainage enterprises is largely in the Corn Belt, the Lake States, and the Mississippi Delta. Between 1920, when the first census of drainage was taken, and 1950, the greatest increase of land in drainage enterprises occurred in the Delta and the Corn Belt states. The southeastern region had a high proportional increase. The relatively little drainage work done in the western states has been mostly in connection with irrigation.

The millions of acres, once too wet to be cultivated and since reclaimed, now rank among the most valuable agricultural areas of the country. Additional millions of acres, where crop losses were frequent, owing to inadequate drainage, now produce well.

In southeastern Missouri one drainage district of 400,000 acres, with other smaller districts, covers much of the agricultural area of five counties. In 1909, before drainage, the only land under cultivation consisted of small patches along streams. Less than 5% of it was cultivated, and the rest was swampland. Drainage work was started about 1912. Now, 95% of the area is under cultivation and includes some of the most valuable land in Missouri.

Drainage has added 50–60 million acres of fertile swampland to the cultivatable areas of the United States, and cultivation has been increased on an additional 75–100 million acres. An estimated 20 million additional acres of fertile undeveloped land needs to be drained if farmland is to be developed from it. Roughly 7 million acres of this drainable land is found in the fertile bottomlands of the Mississippi River, in Arkansas, Louisiana, and Mississippi. Another 7–8 million acres are scattered in the Coastal Plains and other parts of the Southeast. But if this acreage were shown on the map in Figure 13, the dark, dotted space would be increased by about one-fifth; in the

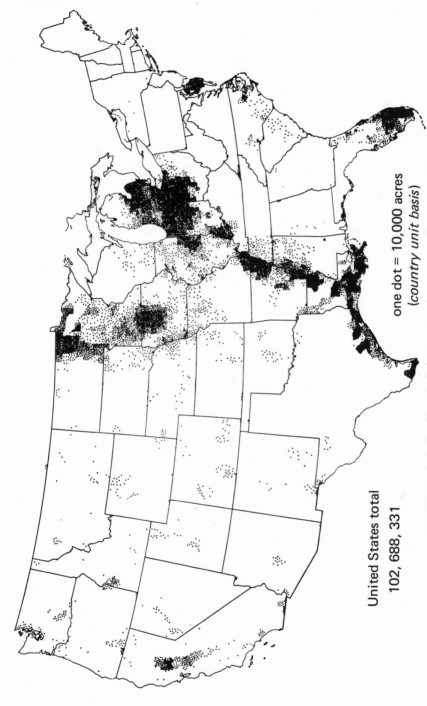

United States total
102, 688, 331

one dot = 10,000 acres
(*country unit basis*)

FIGURE 13. Agricultural land in drainage enterprises, acreage January 1, 1950.

Southeast it would probably be more than doubled. Not all wetland is suitable for drainage. At least 75 million acres of wetland in the United States are unsuited for agriculture under present conditions, but they can be used for wildlife, forests, and recreation. Tables 24 and 25 give statistical data.

The Okefenokee Swamp and the Everglades are areas unsuited to drainage for agriculture. The Okefenokee is an oasis of primitive beauty (in the eyes of a naturalist) and is some 680 square miles in area. It is a realm of stately cypress trees, peat quagmires, and dim waterways. There are sandy pine islands, sunlit prairies, and clear lakes, with alligators in the fastnesses of the swamp, and serpents, a wide variety of amphibians, black bear, little deer, bobcat, and raccoon. Humans are welcome only as visitors: Okefenokee is a national preserve.

Table 24. Acreage of Land in Organized Drainage Enterprises Regions and by Specified Years, 1920–1950

Region[a]	Area in thousands of acres				Change[b] 1940–1950
	1920	1930	1940	1950	
Northern					
Northeastern	0	0	578	744	166
Lake States	19,757	21,548	20,730	21,979	1,249
Corn Belt	28,924	32,700	32,194	35,194	3,000
Northern Plains	2,164	2,929	3,227	3,457	230
Total	50,845	57,177	56,729	61,374	4,645
Southern					
Appalachian	1,265	1,873	1,908	2,750	842
Southeastern	1,843	6,247	6,016	6,506	490
Mississippi Delta	7,347	11,275	11,703	19,886	8,183
Southern Plains	2,178	3,054	4,416	6,096	1,680
Total	12,633	22,449	24,043	35,238	11,195
Western					
Mountain	810	1,970	2,773	2,671	−102[b]
Pacific	1,207	2,812	3,422	3,405	−17[b]
Total	2,017	4,782	6,195	6,076	−119[b]
United States	65,495	84,408	86,967	102,688	15,721

Source: Bureau of the Census: Census of Agriculture, 1950, *Drainage of Agricultural Lands*, vol. 4, 1950.
[a] Including irrigation enterprises that have their own drainage.
[b] Decrease indicated by minus sign.

Table 25. Drainage Development and Use of Land in Drainage Enterprises
in the United States to 1950

	Area in thousands of acres			
Year	Land in all enterprises	Drained	Improved	In planted crops
Before 1870	171	150	133	97
1870–79	428	404	373	288
1880–89	2,429	2,267	2,173	1,865
1890–99	3,743	3,500	3,256	2,482
1900–04	5,769	5,414	5,134	3,814
1905–09	12,192	11,081	10,340	7,652
1910–14	19,573	14,138	12,281	9,006
1915–19	18,012	16,262	14,067	10,268
1920–24	11,272	10,028	8,848	6,480
1925–29	7,411	6,824	6,188	4,511
1930–34	2,093	1,974	1,788	1,326
1935–39	3,874	2,962	2,808	1,824
1940–49	15,721	—	14,749	—
Total to 1950	102,688	—	82,138	—

Source: Bureau of the Census, *Drainage of Agricultural Lands* [*1940*], *1942; Drainage of Agricultural Lands*, vol. 4, 1952.

3–6 GLACIERS

When snowfall exceeds loss by evaporation, the remaining snow compacts into ice by pressure. The resulting mass may descend slowly, as a glacier, or more swiftly in ice falls or avalanches. On steep slopes the foundation is unstable from spring rains and blowing winds, and the mass descends. Avalanches are thus masses of snow mixed with ice, soil, pebbles, and boulders which reach a condition of instability and slide down a mountain slope with destructive force. Thunderstorms, and skiers who disobey the avalanche patrol, may also cause avalanches. Air displacement causes strong winds sufficient to uproot trees adjacent to the course of the avalanche. Everywhere we have learned to live with avalanches and keep them under control. In the Alps, rifle shots, small artillery, and even dynamite are employed to trigger avalanches, thereby removing their menace. They occur in the high mountain areas of the Alps in Europe, the Andes, the Himalayas, the northern United States (Glacier National Park on the Canadian border), and Alaska, as well as in the Arctic and Antarctic, Greenland, Iceland, Siberia, and eastern USSR.

The limit above which snow remains throughout the year is called the snow line, and varies from 15,000–16,000 ft in the tropics and in the Peruvian, Bolivian, and Chilean Andes, to sea level in the Polar regions.

The Alpine types of glacier have been studied for a long time. Owing to the weight of accumulated snow on gentle slopes in the high mountain valleys and on the summits of dome-like peaks, the lower layers are compacted to dense, clear ice with a granular structure known as névé, or firn. The ice is stratified from falls of snow and melting between falls, and from thin layers of dust from the air or the melted snow. Debris interferes little with mountain climbers, but cracks in the ice or crevasses are dangerous. At the head of the clear ice granular structure is a large symmetrical crevasse lying a short distance from the exposed rock surface, where the rock has been disintegrated by frost action. The debris formed falls to the bottom of the crevasse to form a moraine.

Glaciers creep, with the greatest movement being in summer, depending on the size of the mass, the slope and smoothness of the valley floor, and the slope of the upper ice surface. The movement may be compared to a turgid stream, greatest in the center, in the upper layers, and in restricted channels.

In summer, when melting is greater than precipitation, the granular ice appears beneath the snow. This is the firn line, where boulders and debris come to the surface and large boulders sink, but flat rocks protect the ice from melting so that they are perched as in ice tables. Dust and small rocks absorb the sun's heat more quickly, so that the ice beneath them melts and they sink. The moraines increase as the glacier passes to lower altitudes. Finally, the ice sheet melts to water in which the rock burden sinks to form a series of mounds of the terminal moraine. The lower end of a valley glacier may advance and retreat over a period of two decades, depending on variations in climatic conditions, the severity of winters, the heat of summers, and other meteorological conditions.

The Alaskan glaciers consist of great valley glaciers that melt to form a large sheet of very slowly flowing ice. The Malaspina Glacier covers 1500 square miles, and has enough debris to support forests. Precipitation and evaporation loss are very closely in balance.

In Greenland, the greater part of the land is covered by an ice sheet sloping very gently to the coast. Valley glaciers occur only along the coast where rocky peaks protect the ice sheet. The moraines are not large and consist of rocks abraded from the underlying rock structure. Along stretches of the coast the ice forms vertical or overhanging cliffs. The Greenland Icecap is the source of a large number of icebergs during the summer. These drift to lower latitudes and gradually disappear by melting. But it is necessary to operate an iceberg patrol in the shipping lanes to warn of dangerous icebergs.

In some parts of the Himalayas the glaciers are reported to be retreating. Glaciers assisted by frost action on exposed rocks are important eroding mechanisms, resulting in pyramid-shaped mountain peaks, hanging and U-shaped valleys, and truncated spurs and steps in the valley floor from the deepening of valley glaciers. There are many other effects of retreating glaciers: perched blocks, clay beds, lateral and terminal moraines, old lake terraces, and a wide variety of lakes.

3–7 DESERTS

Once used only for uninhabited—"deserted"—places, the term desert is now commonly applied to regions characterized by meager rainfall, scanty vegetation, and limited human use. However, of the three types of deserts commonly recognized, only a few tropical deserts, such as the Libyan part of the Sahara, resemble the popular image of a hot, dry, sandy expanse. Middle-latitude deserts lie in the "rain-shadow" of a mountain barrier (e.g., the deserts in the southwestern United States) or deep within the moisture-starved interiors of continents (e.g., the Gobi). Polar deserts occur because moisture is "locked up" as ice and snow.

Civilizations have attacked the deserts from time immemorial with canals, irrigation, and reclamation. Present-day societies, which number among them all of the technologists, engineers, and scientists who ever breathed, 90% of whom are still living, have been particularly successful in reclaiming deserts, especially the Imperial Valley in the United States, which has been transformed into a major food granary (Table 26).

The Atacama and the Kalahire deserts in Chile are caused by their nearness to the sea, in that they are on a western coast in the lee of prevailing winds and a cold coastal current. The icy Humboldt current, reinforced by the upwelling of deeper water, cools the breezes reaching the high coastal mountains of the Andes, the ascent cooling them more than the land can warm them. The cloud banks and fogs seem to promise rain, but from Arica to Caldera there is less than 1 inch per year. When the Humboldt is displaced by a warm tropical water stream, there are years of disaster. In the years of El Niño (Spanish for "Christ Child at Christmas") torrential rains are let loose on the dust-dry desert hills of the Peruvian Coasts. The soil washes away, the mountain stones roll, the mud and adobe huts disintegrate, and crops are ruined. Their fisheries, among the world's richest, are wrecked when the cold water fauna sickens and dies in warm water. Birds and animals that eat fish to live, die, and wastes are rampant—far beyond human control or influence.

Table 26. Principal Deserts of the World

Desert	Location	Size (approx)	Comments
Atacama	Northern Chile	600 miles long	An area of nitrate and low copper content deposits
Black Rock	Humbolt and Pershing Counties, Nevada, US	1000 square miles	A barren plain, devoid of vegetation, from its surface, alkaline dust is blown into vast clouds by summer winds
Colorado	Arid region of southeastern California	200 miles long, max. width 50 miles	Includes Imperial Valley whose soil is highly productive when irrigated
Dasht-i-Kavir	Iran, from Caspian Sea to Persian Gulf	300 miles long, 100 miles wide	Saline swamps and dry salt areas
Death Valley	Eastern California and southwestern Nevada	2936 square miles	Yielded much borax (20-mule team) in 19th century
Kara Kum	Asiatic USSR	110,000 square miles	Has old river beds of channels or tributaries of the Amu and other rivers
Kyzyl Kum	Kazakhstan, USSR	370 miles × 220 miles	Stationary sands with sparse vegetation
Negev	Southern Israel	4700 square miles	Barren expanses of sand, now irrigated, with farms, orchards, and towns; King Solomon's mines and the Dead Sea, sources of chemicals, are there
Namib	Southwest Africa	800 miles long, 30 to 100 miles wide	Long narrow desert plain along Atlantic Coast, parallel to cold Benguela current of the ocean
Sahara	North Africa, on the west extends to Atlantic Ocean	Over 3,000,000 square miles	Nomadic herding, petroleum exploration and oil production; the Libian and Nubian Deserts included

Name	Location	Size	Notes
Gobi	Extends 1500 mi from Manchuria to Sinkiang province of China	500,000 square miles	Many paleontological finds, including dinosaur eggs
Syrian	Northern Arabia above latitude 30° north	—	Oasis of Jauf has palm groves and 10,000 people
Nufud	Arabia, south of Jauf	Avg. width of 200 miles	Is habitat of Bedouins during spring season
Rub al Khali	Southern Arabia	300,000 square miles	Practically unexplored, the "Empty Quarter"
Great Australian	Much of Central and Western Australia	—	Primitive Bushmen live there; Australians call it the "Sparse Lands"
Kalahari	South Africa, southern limit is Orange River	120,000 square miles	Home of the Hottentots and Bushmen
Mohave	Southern California, NE of Los Angeles, extends into Mohave County, Arizona	15,000 square miles	Needles, on the Arizona–California border, is one of the hottest towns in the US
Painted Desert	Coconino and Navajo Counties in northern Arizona	200 miles long, 15–30 miles wide	Hopi Indian villages are within desert; called "painted" because of coloring, caused by centuries of erosion, exposing brilliantly colored rock formations
Taklamakan	Sinkiang Province, China, Central Asia	125,000 square miles	Ancient trade routes cross, buried cities reported; the Chinese exploded their first atomic bomb there in 1964
Thar	Northwest India, between the Gulf of Cutch and the Arabian Sea on the south	100,000 square miles	Camel caravans still use trade routes through area; parts are in dispute between India and Pakistan

The importance of water to the creation and sustenance of every living thing was profoundly appreciated by the builders of civilization in the ancient Middle East. The vital role of the water cycle, by which the supply of water on the land is constantly replenished, and the loss of an adequate supply, resulted in the demise of many human settlements in the ancient world. Notable was the disappearance of the high civilizations of the "fertile crescent" in the Tigris and Euphrates River Valleys, the Persian Empire, and more recently the fall of the Roman Empire and the decline of the Arabian caliphates, which has been blamed on the more or less sudden decline in the amount of water available. Potentially prosperous regions of Africa, Asia, and Australia, stretching over millions of square miles in the tropical latitudes, presently have only a sparse population because of a lack of water.

In primeval days, the earth was devoid of atmosphere. The earth had no oceans, and its surface was covered with lava-emitting volcanoes, along with hydrogen and hydrogen-rich compounds and water vapor. Some water molecules were split by sunlight to hydrogen and oxygen. The hydrogen escaped, and the oxygen reacted with metals to form oxides and with ammonia and methane to form nitrogen and carbon dioxide. As the earth cooled, plants appeared, absorbing the carbon dioxide and liberating oxygen; water molecules condensed and formed the oceans. The amount of water on the earth can be considered to have been constant during the evolution of man.

Underground seas have been discovered in Iran, Arabia, and North Africa, where ancient peoples used these aquifers some 3000 years ago. They built a system of underground channels (*qanats*) for tapping underground water.[4] The system was expanded under the Islamic caliphates to the Middle East and to North Africa. There were *qanat* networks of underground channels bringing water from the highland aquifers to the low lands by gravity, as is done in Iran today for three-quarters of its water needs.

Arabia found its underground water as a result of drilling for oil. The eastern region has seven aquifers in Cenozoic and Mesozoic strata. The gradient is west–east. The existence of underground water in large quantities under the Sahara has been confirmed. There are seven major storage basins, as shown in Figure 14, having a total area of 4.5 million km^2 and 15.3 trillion m^3 of water. The Western Desert basin of Egypt covers 1.8 million km^2 and has a storage capacity of 6 million m^3.

The Great Eastern Erg of Algeria has an area of 375,000 km^2 and 1.7 trillion ft^3 of fossil water, gathered when the Sahara was a tropical region with heavy rainfall. The aquifer is now charged by flood rains from the Tell Mountains. The discharge occurs by evaporation from the Chott Melrhir-Djerid, which is dry except during the rainy seasons. Its lowest point is 30 m below sea level and extends over an area of 10,000 km^2.

[4]H. E. WULFF, *Scientific American*, April, 1968.

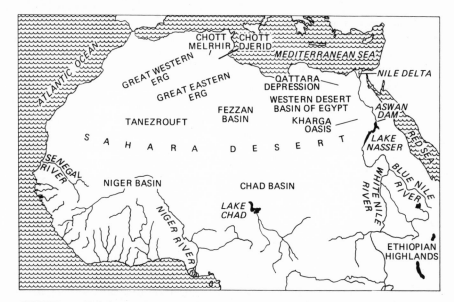

FIGURE 14. The Sahara Desert contains underground water resources concentrated in seven major storage areas. These are the Great Western Erg, the Great Eastern Erg, the Tanezrouft, the Niger, the Fezzan, the Chad, and the Western Desert basins. The total water storage is estimated to be 15.3×10^{12} m³.

Hummerstone[5] has reported that Libya asked Occidental Petroleum, in 1967, to drill for water in the Libyan Sahara, and a subterranean reservoir was discovered. Rainfall is negligible and the temperatures reach 170°F. The aquifer is believed to extend hundreds of miles and to contain the equivalent of the Nile's flow for 200 years.

At Kufra, 300 miles south of the Mediterranean, a somnambulant oasis has 35 fields, each two-thirds of a mile across, and fed by irrigation from the aquifer. Because labor in the area was short, automation was needed. There is a single well in the center of each circular field, each with a diesel engine to pump the water. An overhead pipe 1850 ft long is connected to the wellhead and pivots around it, spraying water on the crop as it turns. Each irrigator is supported on wheels, every 230 ft of pipe. The pipe mechanism is driven by electric motors using current from a diesel-driven generator at speeds of one revolution per 20, 40, or 60 h.

The present 8500 acres are producing four crops per year of wheat and barley and 12 crops per year of alfalfa, with 130,000 acres projected. There are 250 acres in each circular unit. At the predicted rate of use the water table is expected to drop 115 ft in 40 years, or about 3 ft per year.

[5]R. D. HUMMERSTONE, *Fortune*, 1973.

The water cycle over the planet is a distillation extending over the surface. The heating of the subtropical regions by solar radiation leads to a continuous evaporation of water to the atmosphere, removal by wind currents, and transportation—mostly in the lower half of the atmosphere—northward and southward into other latitudes.

During the process of transfer in the gaseous phase, some vapor condenses to form clouds because of cooling by expansion. These condense and precipitate over the equatorial, middle, and high latitudes. Clouds, rain, typhoons, rivers, and underground flow return the water to the oceans.

3–8 RECLAMATION

As the conservation arm of the Federal Government, the Department of the Interior defines its task in terms of the entire planet. From recreation to reclamation, from minerals to mallards, from water to wilderness, the various programs within the department must meet specific problems and still mesh with other efforts—efforts to preserve our historic past, to stretch our natural resource potential, and to restore, beautify, and enhance our overall environment.

Originally designated the Reclamation Service, the agency was created 6 decades ago to irrigate the dry lands of the West and make them capable of settlement. By building those arid acres into a viable economy, reclamation served not only to benefit that area but the entire nation.

As the water needs of America grew, the role of the bureau expanded. Bureau functions multiplied to include irrigation, hydropower generation, flood control, municipal and industrial water supply, water-quality improvement, outdoor recreation, and fish and wildlife enhancement.

Its whole method of operating changed as the ratio between human needs and natural resource supplies narrowed. More people require more food, more electricity, more water to drink and use in their homes, more water to work in their industries, and more recreation.

Gradually, the bureau became vitally concerned with all sources of water—water in natural waterways, ground water, moisture carried in the atmosphere, and the sea. In its early days it usually built only a single-purpose project on one site. Its operations presently include development of the entire length of a stream, then of a whole river basin. It is engaged now in interbasin development as the most efficient and most economical method of filling the nation's urgent need for water.

The regional approach to water resources is embodied in the Colorado River Basin project of the Bureau of Reclamation and others to solve the serious water-supply problems of a large section of the arid West. Benefits have included irrigation, municipal and industrial supply, flood control, hydroelectric generation, water quality regulation, recreation, and fish and wildlife betterment. The comprehensive plan provides for construction of dams, power plants, and distribution systems to meet the needs of the basin area. The Colorado River Basin project is the joint creation of the states, the local entities of the area, and the Federal Government.

The 1966 period saw completion of 11 dams with reservoirs having a combined impoundment capacity of 2.8 million acre-feet of water. (One acre-foot equals approximately 326,000 gal.) A total of 457,200 kW of electric power generating capacity was installed, as were 900 miles of high-voltage transmission lines. Construction of 733 miles of canals, pipelines, laterals, and drains was also completed. These additions brought the number of reclamation dams to 248, capable of impounding 129.5 million acre-feet of water, enough to put New York State under nearly 4 ft of water. At the end of 1966, 17 dams were under construction, having a total reservoir capacity of 5 million acre-feet, 4 power plants, and 2 pump-generating plants, with a combined capacity of 917,200 kW, and with 623 miles of high-voltage transmission lines.

Ground water overdraft by pumping from the Arvin–Edison Water Storage District in the San Joaquin Valley has been averaging 200,000 acre-feet (nearly 65 billion gal) a year. A distribution system has now been constructed to convey surface water from the bureau's Friant–Kern Canal and will deliver enough water to halt the overdraft. Much of this water will be allowed to percolate into the soil to replenish the ground water aquifer, and some will be used for surface irrigation.

During 1966, the first 16-mile section of the San Luis Canal in California was completed and work began on the final reach of what will be one of the largest artificial rivers in America, capable of carrying about 100,000 gal of water per second. The concrete-lined canal conveys water southward 103 miles from the San Luis Dam near Los Banos to irrigate one-half million acres of land. The State will use it to carry over 1 million acre-feet of water annually for irrigation and municipal and industrial use, transporting the water by a state-owned canal still farther south across the Tehachapi Mountains to the southern California coastal area.

The San Luis Dam, completed in 1967, impounds more than 2 million acre-feet of water. The entire system, completed in 1970, has over 1000 miles of irrigation pipelines. Its costs are to be repaid by the water users under a $157,048,000 contract with the Westlands Water District—the largest such contract in the history of reclamation. Water for this distribution system is imported from northern California via facilities of the Central Valley project, with structures of the San Luis Unit providing the final link.

Within the upper Colorado Basin project, the river has been developed in a 110,500-square-mile area, embracing parts of Colorado, Utah, Wyoming, New Mexico, and Arizona.

On the Curecanti Unit in southeastern Colorado, Blue Mesa Dam was constructed, a 340-ft-high earthfill structure, with a 60,000 kW power plant, plugging the Gunnison River. By 1967, 465-ft-high Morrow Point Dam, 12 miles downstream and with an underground 120,000-kW power plant, was completed.

Because of the mountainous terrain of the Colorado River Basin, much of the reclamation work there required tunneling to divert water from one watershed to another. In 1905, the Bureau started a tunnel to transport Gunnison River water through a rocky ridge to the Uncompahgre Valley in Colorado.

Reclamation's development of the 10-state Missouri River Basin project began with the Garrison Diversion Unit. Water from the Garrison Reservoir behind Garrison Dam, on the mainstem of the Missouri River in central North Dakota, provides irrigation for 250,000 acres in the state.

Since 1967, the Canadian River project in Texas has supplied 11 cities in the panhandle region of that state. The project includes Sanford Dam on the Canadian River and a 322-mile pipeline distribution system.

A fundamental feature of the reclamation program is that, to a large extent, the direct beneficiaries of a project pay for it. Nearly 90% of the cost of constructing multimillion-dollar developments to conserve water for the use and enjoyment of all the people is repaid to the U.S. Treasury in cash. Federal investment for power and for municipal and industrial water facilities is repaid, with interest, by electricity and water users, and that for irrigation features is repaid in full, but without interest. About one-half of the cost of irrigation elements is borne by the water users and the remainder is repaid through the sale of surplus power and from other revenues after costs allocated to power and additional reimbursable purposes have been met.

National recreation areas are at reclamation reservoirs, examples of which are: Bighorn Canyon, in Montana and Wyoming: Whiskeytown–Shasta–Trinity, in California; Rocky Mountain, in Colorado; Lake Mead, in Arizona and California; and Coulee Dam, in Washington.

Because much of the water in the western states comes from the melting of winter snow-pack in the mountains, the bureau's primary research during the year concentrates on precipitation induction (cloud seeding) of mountain-cloud systems. Since summer thunderstorms are a major source of moisture for the Great Plains, the researchers also study the characteristics of convective cloud (thunderstorm) systems.

One facet of weather modification research is the development of instruments to transmit meteorological and hydrological information from remote or inaccessible sites to a central point where the facts can be analyzed. Aircraft equipped with intricate instruments obtain needed data from within the clouds.

The resource needs of underdeveloped countries are virtually identical to those of the United States. However, they generally are more immediate and intense. Such nations lack the knowledge and training necessary for creating and operating water-supply systems for agricultural, human, and industrial consumption, and for power generation.

Ladurie[6] traces the history of climate since the year A.D. 1000. Olive growing moved northward from 1550 to 1600 because growers were seeking expanding markets, but actually it was a time of colder climate. Ladurie states that late medieval reductions in grape growing resulted from plagues and the disruptions of war, causing high labor costs.

In France and Germany, early harvests were the result of warm springs and summers associated with anticyclonic conditions. Late harvests resulted from cool springs and summers with cloudy cyclonic periods. Vintage dates and wine quality were related to meteorological conditions. Ladurie records that Germany had warm summers from 1453 to 1552, and that from 1552 to 1600 weather was unfavorable for wine production.

That man lives and is dependent on water on our water planet is illustrated in the 1970 report of Estimated Use of Water in the United States, which is updated every five years.[7]

Water use in the United States in 1970 averaged about 370 bgd (billion gallons per day), or about 1800 gal per capita per day. This amount was withdrawn for the four principal off-channel uses, which are (1) public supplies (for domestic, commercial, and industrial uses), (2) rural supplies (domestic and livestock), (3) irrigation, and (4) self-supplied industry (including thermoelectric power). In 1970, withdrawals exceeded the 310 bgd estimated for 1965 by 19%. Increases were, approximately: 25% for self-supplied industry (mainly in electric utility thermoelectric plants), 13% for public supplies, 13% for rural supplies, and 8% for irrigation. Industrial water withdrawals included 54 bgd of saline water, a 20% increase in 5 years. Hydroelectric power, (an in-channel use) amounted to 2800 bgd, a 5-year increase of 22%. The quantity of fresh water consumed—i.e., water made unavailable for further possible withdrawal because of evaporation, incorporation in crops and manufactured products, and other causes—was estimated to average 87 bgd for 1970, an increase of about 12% since 1965.

Estimates of water withdrawn from the principal sources indicated that 68 bgd came from fresh ground water, 1 bgd came from saline ground water, 250 bgd came from fresh surface water, 53 bgd came from saline surface water, and 0.5 bgd was reclaimed sewage.

The average annual streamflow—a simplified measure of the total available water supply—is approximately 1200 bgd in the coterminous United States. Total water withdrawn in 1970 for off-channel uses (withdrawals other

[6]E. L. LADURIE, *Times of Feast, Times of Famine*, translated from French by Barbara Bray, Doubleday, Garden City, New York, 1971, 426 pp.

[7]*U.S. Geological Survey Circular* 676, U.S. Gov't Printing Office, Washington, D.C.

than for hydroelectric power) amounted to about 30% of the average annual streamflow; 7% of the 1200 bgd basic supply was consumed. However, comparisons of Water Resources Council regions indicate that the rate of withdrawal was higher than the locally dependable supply in the Middle Atlantic, Texas–Gulf, Rio Grande, Lower Colorado, and California–South Pacific regions. Consumption amounted to nearly 25% of withdrawals in the coterminous United States; however, fresh water consumption amounted to only 14% of off-channel withdrawals in the 31 eastern states and ranged from 30% to nearly 70% of off-channel withdrawals in the Water Resources Council regions in the West. In the Rio Grande and Lower Colorado regions, freshwater consumption in 1970 exceeded the estimated dependable supply of fresh water.

Interpreting from Figure 15, the per capita use in potable water is of the order of 180 gal per day, and for all uses 1800 gal per person per day. (See also Figures 16 and 17; Tables 27 and 28.)

3-9 GEYSERS

In many places, faults in the earth's crust have allowed molten magma from the planet's interior to intrude near the surface. In some of these areas, ground water percolating down into the earth comes into contact with the hot solidifying magma. If the heated water rises to the surface, it is manifested as hot springs, geysers, or fumaroles. Wells have been drilled in these to produce steam or hot water for the generation of electric energy. The world's first and largest geothermal power plant is near Lardarello, Italy.[8] There, 380,000 kW of geothermal power are generated using steam from a vapor-dominated geothermal reservoir.[9] On New Zealand's North Island, a 175,000-kW generating capacity is installed using the energy of hot water from the Wairakei geothermal field.[10] A third facility, a 75,000-kW power plant being built near Mexicali, Baja California, Mexico, will use steam flushed from hot brine that has about the same salinity as seawater.

Finney[11], of Pacific Gas and Electric Company, has described the Geysers geothermal power plant on a resort site featuring mineral baths. It is now the only United States geothermal plant, which by 1974 became the largest of its kind in the world. Located in a mountainous area some 80 miles north of San Francisco, its six turbine generator units have a total net capacity of

[8] J. BARNEA, Scientific American, 226: 70 (1972).
[9] D. E. WHITE, L. J. P. MUFFLER, and A. H. TRUESDELL, Economic Geology 1: 66, 75 (1971).
[10] Power from the earth: The story of the Wairakei Geothermal Project. New Zealand Ministry of Works, 1970.
[11] J. P. FINNEY, Chemical Engineering Progress, 68: 83–86 (1972).

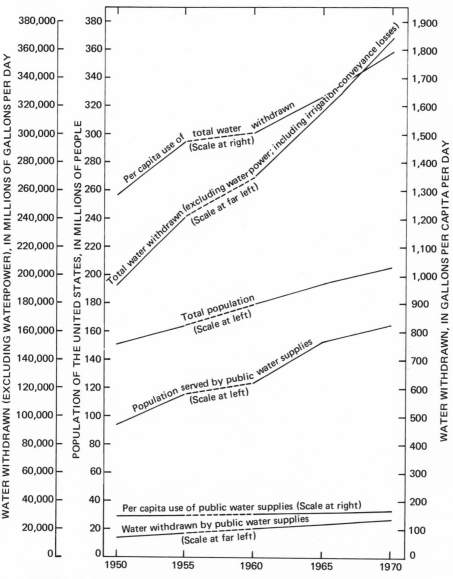

FIGURE 15. Trends in population and withdrawals of water in the United States, 1950–70.

Table 27. Changes in Water Withdrawals and Consumption in the United States
(billion gallons per day, 1950–70)

	1950	1955	1960	1965	1970	Percent change 1965–70
Total population (millions)	150.7	164	179.3	193.8	203.2	5
Public supplies	14	17	21	24	27	13
Rural domestic and livestock	3.6	3.6	3.6	4.0	4.5	13
Irrigation	110[a]	110	110	120	130	8
Thermoelectric power (electric utility) use	40[b]	72	100	130	170	33
Other self-supplied industrial use	37[b]	39	38	46	47	2
Total withdrawals	200	240	270	310	370	19
Fresh ground water	34	47	50	60	68	13
Saline ground water	—	0.65	0.38	0.47	1.0	113
Fresh surface water	160[c]	180	190	210	250	19
Saline surface water	10[c]	18	31	43	53	23
Reclaimed sewage	—	0.2	0.1	0.7	0.5	−29
Water consumed by off-channel uses	—	—	61	77	87	13
Water used for hydroelectric power	1100	1500	2000	2300	2800	22

[a]Including an estimated 30 bgd in irrigation conveyance losses.
[b]Estimated distribution of 77 bgd reported by MacKichan (1951).
[c]Distribution of 170 bgd of fresh and saline water reported by MacKichan (1951).

Table 28. United States Self-Supplied and Irrigation Water Withdrawals
(million gallons per day)

State	Self-supplied industrial water withdrawals	Irrigation water withdrawals
Alabama	6,100	18
Alaska	180	0.9
Arizona	220	6,300
Arkansas	1,500	1,300
California	11,000	33,000
Colorado	320	13,000
Connecticut	3,100	5.9
Delaware	5,200	76
Florida	12,000	2,200
Georgia	4,600	47
Hawaii	1,300	1,300
Idaho	450	15,000
Illinois	13,000	21
Indiana	8,000	25
Iowa	1,700	26
Kansas	420	3,000
Kentucky	4,200	7.1
Louisiana	7,000	1,600
Maine	620	8.9
Maryland	4,600	6.6
Massachusetts	3,400	58
Michigan	11,000	58
Minnesota	2,900	20
Mississippi	1,500	370
Missouri	2,800	77
Montana	210	7,600
Nebraska	990	4,700
Nevada	140	3,000
New Hampshire	600	2.8
New Jersey	5,200	76
New Mexico	110	2,800
New York	15,000	27
North Carolina	5,200	82
North Dakota	370	190
Ohio	17,000	31
Oklahoma	360	820
Oregon	720	4,800
Pennsylvania	18,000	10
Rhode Island	350	4.5
South Carolina	3,000	29
South Dakota	190	230
Tennessee	5,900	4.2
Texas	15,000	10,000
Utah	290	3,600
Vermont	51	0.1
Virginia	5,000	35
Washington	600	5,600
West Virginia	5,600	1.3
Wisconsin	5,600	52
Wyoming	310	5,400

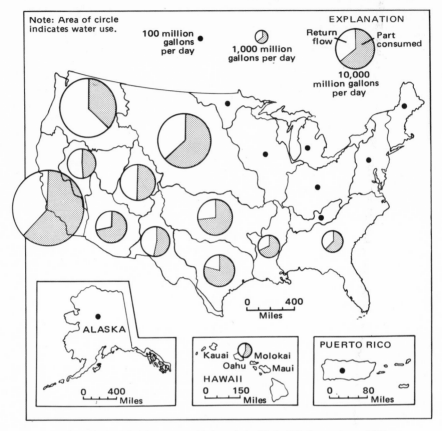

FIGURE 16. Irrigation water withdrawals by United States regions, 1970.

184,000 kW. These units use steam directly from a geothermal dry-steam reservoir similar to the Lardarello steam field. The Geysers area extends along a steep, narrow canyon of Big Sulfur Creek, a tributary of the Russian River. As the name implies, the area has long been noted for its natural steam vents, or fumaroles, hot mud pots, and brilliantly colored rock formations altered from geothermal activity.

In the 1860s, a resort was built to take advantage of the Geysers' natural steam vents for hot mineral baths. In the early 1920s a group of area businessmen formed the Geysers Development Company to try to develop electric power. Eight shallow steam wells were drilled. They were productive, but at that time undeveloped hydroelectric power potential was lower in cost and so plentiful in California that power production at the Geysers was uneconomical.

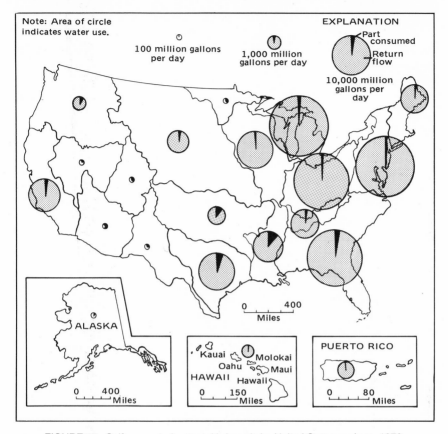

FIGURE 17. Self-supported water withdrawals by United States regions, 1970.

Two small reciprocating steam engine sets installed for lighting at the resort became the first authentic geothermal power plant in the United States.

In 1955, the Magma Power Company obtained leases on most of the property around the original steam wells. Engineering studies followed which showed that a small turbine generator unit could be installed and operated economically if the steam price was about 2.5 mills per kWh. There are now 11 widely distributed units in a rectangle measuring 4 miles × 2 miles. The generation of electric power has proven commercially successful at the Geysers. Power production costs there compare favorably with those of PG&Es latest supercritical fossil units.

Attention to design, operation, and maintenance has overcome problems associated with particulate matter in the steam and the corrosiveness of the condensate. Additional generating capacity is scheduled each year to total 502,000 kW, making the Geysers power plant the largest geothermal power installation in the world. Beyond that, additional capacity will continue to be installed as additional steam reserves are developed.

4

THE ATMOSPHERE

4–1 COMPOSITION

The atmosphere consists almost entirely of nitrogen, oxygen, and argon. A fourth major component, carbon dioxide, essential to all plant life, is present only to the extent of about 300 ppm. The average composition is given in Table 29, based primarily on the work of Paneth.[1]

Convection currents rising to 20 km (12–13 miles) above sea level, ceaselessly mix the air, and thereby keep the analysis reasonably constant. Gravitational separation does not occur until about 150 km (80 miles). Above 600 km, the atoms and molecules describe elliptical orbits in the earth's gravitational field.

As the barometric pressure decreases and elevation above sea level increases, the need for human adaptation grows more pressing. Many of my engineer friends and associates are uncomfortable at Bogota, Colombia, 8563 ft high, and others need periods of acclimatization at La Paz, Bolivia, 12,500 ft high. Those who are born and live on the Bolivian altoplano develop oversized lungs and are pouter chested. When they descend to work in the much lower agricultural regions, they are highly susceptible to pulmonary diseases and often die within a year from lung atrophy.

The lower portion of the atmosphere, where convection occurs, is the troposphere. This is the region in contact with the earth's surface, the realm of clouds, rain, snow, hurricanes, and tornadoes. The troposphere is about 17 km (11 miles) at the equator and 6–8 km (4–5 miles) at the poles. In the troposphere temperatures decrease with decreasing pressure or atmosphere. Above this, in the stratosphere, stratification is evident and an artificial atmosphere must be supplied, via space suits, oxygen supply, and lowered exertion.

---

[1] F. A. PANETH, Atmospheric Composition, *Quart. J. Roy. Meteorol. Soc., 63:*433 (1937).

Table 29. Composition of the Atmosphere

Gas	Composition by volume (ppm)	Composition by weight (ppm)
Nitrogen	780,900	755,100
Oxygen	209,500	231,500
Argon	9,300	12,800
Carbon dioxide	300	460
Neon	18	12.5
Helium	5.2	0.72
Methane	2.2	1.2
Krypton	1.0	2.9
Nitrogen oxide	1.0	1.5
Hydrogen	0.5	0.03
Xenon	0.08	0.36
Ozone	0.01	0.36

Above 50 km is the mesosphere, where temperatures decrease with an increase in altitude, to approximately $-100°C$ at 85 km. Above 80 km, the atmosphere is ionized by the sun's ultraviolet radiation, and the gases become conducting. This is the ionosphere, which reflects back our radio signal waves. The E layer is named after the physicist, Heaviside. F_1 and F_2 are the layers which are absorptive to Hertzian waves.

The atmosphere of the earth extends well beyond 400 km (250 miles). Observation shows the presence of luminous rays up to altitudes of 1000 km (600 miles). Shortwave-radio transmission has shown that the medium is sufficiently dense at several hundred kilometers to produce enough electric charges or electrons to reflect radio waves.

Rocket probes and the drag of artificial satellites at heights of several thousand kilometers show that the terrestial atmosphere extends to great distances.

Carbon dioxide has a highly important function for man, despite its concentration of only a few hundredths of a percent, in that it provides the raw material for plant life. Ozone in the ozone layer of the stratosphere is responsible for absorption of ultraviolet light. If this light were to fall directly on plants and animals, it would prove fatal to most forms of life. The ozone sphere also reduces the escape of terrestrial heat by radiation to space.

The mass of the atmosphere has been estimated by many geophysicists to be on the order of 25×10^{14} metric tons, or approximately 0.00009% of the mass of the whole earth. Our continued existence is therefore absolutely dependent on what are no more than trace constituents of carbon dioxide and ozone (0.01 ppm) on this fragile water planet.

Geologists do not agree how the planet came into being, but are in accord that "many changes must therefore have occurred to produce an atmosphere such as we have today, yet the evidence of the rocks suggests that atmospheric conditions have not changed radically through most of geological time."[2]

Over recorded time, similar glaciations have occurred repeatedly. Rocks show a state of oxidation from the earliest to the most recent times; sedimentary rocks also share this characteristic throughout the geologic eons. The atmosphere seems to have always possessed the same changing character as today, and our presence has apparently not affected its constitution in the slightest.

Today some believe that the earth's environment, though in a beautiful and stable state before our arrival, has been irreversibly altered for the worse. But since the environment of this planet earth has clearly been continuously changing, in state and arrangement of materials, though not in fundamental composition or underlying process, for about 4½ billion years, the earth never has been in a stable state. It has changed and evolved continually and radically through these many eons. In fact, nature abhors stability and loves change.

The atmosphere of the primitive earth probably consisted largely of methane and carbon dioxide. Because of the carbon dioxide, the "greenhouse effect" was very important. The greenhouse effect is a heating of the earth owing to the atmosphere, whose action is similar to that of the glass planes of a greenhouse. The surface of the earth was very warm and the oceans resembled warm soup.

The early forms of life took nature as they found it, and what they found, among other things, was a plentiful supply of carbon dioxide and sunshine. Through a complex system of molecular reactions, the new life forms used sunshine and carbon dioxide to supply the energy and materials they required in order to grow and to multiply.

However, in using carbon dioxide these forms of life changed the atmosphere because they discharged into it an impurity not heretofore present: oxygen. Over several billion years the carbon dioxide content of the atmosphere was drastically reduced and the concentration of oxygen steadily increased. Early forms of life had upset the so-called "balance of nature."

Soon a new kind of life emerged that used the oxygen and discharged carbon dioxide, gaining its energy through oxidation rather than through a photosynthetic process. The human being is the latest in a long line of life forms that take in oxygen and produce carbon dioxide as a waste product. We do know that the oxygen content of the atmosphere has not changed detectably in the past 50 years.

[2]B. MASON, *Principles of Geochemistry*, J. Wiley & Sons, New York, 1952, p. 183.

4-2 WEATHER

Meteorology, the study of the atmosphere (weather) and the science of things in the air, had a technical society in 1859, the Royal Meteorological Society of London. There now exist Scottish, Austrian, German, French, Italian, Spanish, Japanese, Belgian meteorological societies and associated meteorological journals, as well as the American Meteorological Society which has published its *Bulletin* since 1920.

The study of weather goes back to the *Meteorologica* of Aristotle, the Greek, 384–322 B.C., long before the invention of the thermometer (Galileo, 1607); the barometer (Torricelli, 1640); Boyle's gas laws (1659); the hydrometer for moisture in the air (LeSaussare, 1740–1799); the discovery of oxygen (Lavoisier, 1783); and Dalton's 1880 laws concerning the pressure of water vapor in air (partial pressure). Espy published the *Philosophy of Storms* in 1841. By midcentury, there was a network of observation stations in many European countries, the first weather bureaus.

Weather satellites above the earth today track nearly 100 small disturbances per year. Typical of these are dust clouds, leaving Africa, and rain squalls originating in the heated atmosphere near the equator over the tropical Atlantic. About ten will accumulate enough energy to grow into tropical storms. Five or six will enlarge until they are hundreds of kilometers across, with wind intensities of 33–90 m/s² (73–200 mph), storms the meteorologist calls hurricanes.

4-3 HURRICANES

Hurricanes have exacted an annual toll of 50 deaths and \$500 million property damage each year, from 1960 on. Civilization continues to invade vulnerable coastal areas, and to build more homes, cottages, resorts, and marinas, thus increasing human exposure to the destructiveness of hurricanes. Satellite monitoring of hurricane-forming regions makes it easier to locate them and follow their movements. However, predicting the path of a hurricane remains only guesswork at best, scarcely a science. Modifying the peak winds of hurricanes is still an uncertain hope.

Warnings to United States coastal residents and shipping are presently based on predictions of the Miami Hurricane Center of the National Weather Service. The predictions of the path of a particular storm rely on statistical methods, climatological inferences from past collections of data, and the study of the paths of earlier hurricanes. Forecasters are often required to choose between conflicting predictions by different methods, and much depends on the experience of the forecasters. For Atlantic hurricanes, the average error is about 180 km for a 24-hour forecast and 390 km for a 48-hour forecast. Further north, the errors are greater for the same periods.

There is considerable confusion among meteorologists as to hurricane theory, mathematical computer modeling, and weather and climate modification; there is also a lack of useful operational data. The first attempt to seed a hurricane, in 1947, was followed by disaster. A storm that had passed over Florida and was headed east into the Atlantic, when seeded by scientists (the General Electric Research group), struck the coast with renewed fury after abruptly turning around.

In subsequent efforts to modify hurricanes under project Stormfury, a cooperative US Navy–National Oceanic and Atmospheric Administration (NOAA) program begun in 1962, only three suitable hurricanes have occurred within reach of the project's airplanes.

The most important feature of a hurricane is the ring of towering clouds around the eye, or center, of the storm. These clouds generate the heat and ultimately the pressure differences that drive the storm's winds. Stormfury experiments were originally based on the hypothesis that seeding within the eyewall clouds would change the distribution of atmospheric pressure within the storm and reduce maximum winds by 10–15%. Some hurricane researchers now believe that seeding within the eyewall clouds would, if anything, increase the storm's intensity.

Hurricanes are usually violent wind storms in the West Indies. The term corresponds to 12 on the Beaufort Scale and is now used to describe similar storms in other regions, except those in the East Indies and the Chinese Seas, where they are still known as typhoons. Hurricane is from the Carib word *huracan*, imported by the Portugese explorers of the fifteenth century.

The Beaufort Scale is a series of numbers arranged by Admiral Sir Francis Beaufort (1774–1856) in 1806, to indicate the strength of the wind from a calm force, 0, to a hurricane force, 12, defined by him as "that which no canvas could withstand." The British Admiralty accepted the scale for the open sea in 1838, and it was adopted, in 1874, by the International Meteorological Committee for international use in weather telegraphy (Table 30).

Throughout history, hurricanes have annually battered the southern coasts of the United States and the Caribbean seacoast of Mexico and Central America. Their direction is generally northward so that they have hit all areas of the Atlantic coastal areas. The major United States hurricanes since 1900 are given in Table 31. Usually they are destructive because their winds, squalls, and heavy rainfall that floods the shoreline and the inland areas and disrupts communication, travel, and other activities of man.

Hurricane Camille, the most deadly and destructive storm to strike the United States in years—perhaps in history—battered the Gulf Coast shores of

Table 30. Beaufort Scale

Beaufort scale	Corresponding wind	Limits of hourly velocity (statute mph)
0	Calm	Under 2
1–3	Light breeze	2–12
4–5	Moderate wind	12–23
6–7	Strong wind	24–37
8–9	Gale	38–55
10–11	Storm	50–75
12	Hurricane	Above 75

Louisiana and Mississippi the night of September 17–18, 1969. With torrential rains, record winds of 190 mph, and storm surges 20 ft high, Camille laid waste to a 60-mile stretch of coastline. In less than 5 hours the killer had taken more than 300 lives, injured hundreds of people, and wrecked $1 billion worth of property. Camille moved northward to cause flooding that took 108 lives in Virginia and West Virginia before its force was spent.

According to the National Hurricane Center in Miami, "Camille was the greatest storm of any kind that has ever affected this nation." Whole towns were almost blown off the map. Buildings were crushed like matchboxes. Giant ships were hurled up on the beach like toys. Many lives and much property could have been spared if more people had heeded warnings to evacuate and if more extensive seawalls had been built. The saddest lesson of Camille was that there is very little man can do in the face of a storm of such enormous intensity.

Although hurricanes are less devastating than major earthquakes, a single hurricane killed an estimated 200,000 persons in Bangladesh (formerly East Pakistan) in 1970. These storms are the most destructive forms of all atmospheric phenomena.

4–4 CYCLONES

The word "cyclone" was first coined by H. Piddington in the *Sailors' Horn Book* (1855) for tropical revolving storms, to indicate the combined circular and centripetal movement which was once thought to be characteristic of all central systems of low pressure. It is now used not only for tropical revolving storms where the pressure is lowest at the center, but also for typhoons. The winds in consequence tend to blow toward the center but are diverted by the earth's rotation and circle around the center, in a counterclock-

Table 31. Major United States Hurricanes Since 1900

(Courtesy, US Weather Bureau)

Date	Name and Area	Deaths	Estimated damage (millions of dollars)	Comments
8/27 to 9/15/1900	Texas	600	$ 25	Most damage and loss of life caused by storm tide that inundated Galveston Island
9/14–21/1909	Louisiana and Mississippi	350	5	Wide area of Louisiana coast inundated
8/5–25/1915	Texas and Louisiana	275	50	Storm tide of 12 ft inundated Galveston to a depth of 5–6 ft
9/22 to 10/1/1915	Middle Gulf Coast	275	13	90% of buildings destroyed over a large area of Louisiana south of New Orleans
9/2–15/1919	Florida, Louisiana, and Texas	287	22	Severe damage in Louisiana and Texas; over 500 casualties in ships lost at sea
9/11–22/1926	Florida and Alabama	243	73	Very severe in Miami area, and from Pensacola into southern Alabama
9/6–20/1928	Southern Florida	1836	25	Wind-driven waters of Lake Okeechobee over-flowed into populated areas
8/29 to 9/10/1935	Southern Florida	408	6	Barometer 26.35 inches on Long Key; record low in Western Hemisphere
9/10–22/1938	Long Island, NY, and southern New England	600	306	Very heavy wind and storm damage in most of New England
8/7–21/1955	Diane—North Carolina to New England	184	832	Heavy rainfall with near maximum runoff caused severe floods throughout the Northeast; damage exceeded any prior storm on record
6/25–28/1957	Audrey—Texas to Alabama	390	38	Storm-surge over 12 ft caused inundation of Louisiana coast as far as 12 miles inland; offshore installations heavily damaged
6/17–23/1972	Agnes—Florida, Eastern states to Pennsylvania and New York	122	1470	$3 billion in flood damage; the costliest storm in U.S. history

Table 32. World Cyclones, Hurricanes, and Typhoons Outside of the
United States

Date	Location	Deaths (thousands)	Comment
9/26–27/1703	England	8	Hurricane
10/7/1731	Calcutta, India	300	Storm surge at mouth of Hooghly River
10/5/1864	Calcutta, India	50	Much of city stripped by cyclone
10/31/1876	Bakarganj, India	200	Storm surge inundated city
10/8/1881	Indochina	300	Typhoon and storm surge
6/5/1882	Bombay, India	100	Cyclone and storm surge
9/18/1906	Hong Kong	10	Typhoon
10/16/1940	Bengal, India	40	Cyclone
10/10–31/1960	East Pakistan	14	Two cyclones struck Bay of Bengal area; storm-surges followed each cyclone
5/28–29/1963	East Pakistan	22	Cyclone and storm-surges left 1 million people homeless
5/12/1965	Barisal district, East Pakistan	13	Cyclone and storm surge left 5–7 million people homeless
12/15/1965	Karachi, Pakistan	25	Cyclone and storm surge
11/12–13/1970	East Pakistan	300	Cyclone and storm surge devastated coastal areas
10/29–30/1971	India	10	Cyclone and tidal wave struck Orissa state

wise direction in the Northern Hemisphere and the reverse in the Southern Hemisphere. The whole system moves in a fashion usually associated with great wind drifts. Well-formed cyclones are usually accompanied by gales and bad weather. On weather charts they are the lows, or the "depressions," that is, low readings on the barometer, an instrument for measuring atmospheric pressure.

The historical record of world cyclones for nearly three centuries is given in Table 32. The destruction wrought by world cyclones, associated hurricanes, and typhoons, has directly affected several million people, and more than that number of domestic and wild animals. In the face of these storms we flee, if we can, from our habitat. Tell us that we should not live in such areas, and we ask "Where can we go?"

4–5 TORNADOES

The name "tornado" was first applied to a violent squall blowing outward from the front of a thunderstorm on the Gold Coast of Africa. It is now used for small-diameter revolving storms, which are especially frequent in the Mississippi basin, but are also experienced in Australia, Europe, and other places, usually in a less violent form.

They normally occur in the spring and early summer in association with cyclones. The most violent tornadoes are always accompanied by a tornado cloud above that, when fully developed, tapers downward to the earth. A tornado has an advancing movement of 20–50 mph, and its narrow path is strewn with wreckage. The storm is of short duration; it passes in a minute or so, and often runs its course in less than an hour. Barometric pressure in the vicinity of a tornado falls very rapidly. The destructive power is not in the straight winds, but in their counterclockwise, rotating mass, which always moves in an easterly and generally in a northeasterly direction. Tornadoes are immensely destructive of human work, but cause relatively few deaths. The major tornadoes in the United States are listed in Table 33.

In August, 1973, a tornado smashed through West Stockbridge, Massachusetts, killing at least two persons and injuring as many as 40 more. The twister leveled a truck stop near the Massachusetts Turnpike and damaged six houses and a motel in the town. Five miles away, a tornado smashed several buildings and injured a young boy.

Black clouds boiled up into the afternoon sky over eastern Iowa one day in September, 1972, and shortly before 2 P.M. a tornado touched down northeast

Table 33. Major United States Tornadoes Since 1900
(Courtesy, U.S. Weather Bureau)

Date	Location	Deaths	Injuries	Property damage (millions of dollars)
4/7/1908	Lamar and Wayne Counties, Mississippi	100	649	0.9
5/26/1917	Mattoon and Charlestown, Illinois	101	638	2.5
3/18/1925	Missouri, Illinois, and Indiana	689	1,980	17
3/21/1932	Alabama (series of tornadoes)	268	1,874	5
6/25/1944	Ravenna, Ohio; Pennsylvania, West Virginia, and Maryland	150	867	4.2
4/12/1945	Oklahoma and Arkansas	102	689	4
3/21–22/1952	Arkansas, Missouri, and Tennessee (series of tornadoes)	208	1,154	14
5/11/1953	Waco, Texas	114	597	41
6/8/1953	Flint to Lakeport, Michigan	116	867	19
4/11/1965	Illinois, Michigan, Indiana, and Ohio (series of tornadoes)	226	3,000	620
2/21/1971	Mississippi, Louisiana, and Tennessee (series of tornadoes)	121	1,600	19

of Cedar Rapids. It plowed through the countryside for 66 miles, staying on the ground six times longer than the average tornado and left a mile-wide swath of destruction. A mild winter and a lack of heavy snows had earlier left the ground warmer than normal, allowing rapid heating of low-level air. Meteorologists liken this effect on the atmosphere to the boiling produced by heating the bottom of a pan of water. The warm surface air ''boils'' up into the severe thunderstorms that spawn tornadoes.

No one was killed and only three persons were injured. This was partly sheer good fortune, because the storm missed most of the area's small communities. An accurate forecast by the National Weather Service had warned that twisters would hit, and the warning was accurate to within 10 min.

The small size of tornadoes (usually no more than a quarter-mile wide) and their short lives (an average of 20 min) make them extremely hard to study, let alone predict. Research is providing not only a better idea of how tornadoes behave, but also new tools for telling when and where they will strike. Four out of ten National Weather Service tornado predictions are correct.

Tornadoes are much smaller and shorter-lived than hurricanes, which can be up to 400 miles wide and last a week or more. The winds in both storms whirl counterclockwise, but a tornado generates winds as high as 300 mph, about double the speed of those in a hurricane.

During the past fifty years, tornadoes have been bigger killers than hurricanes, claiming nearly 9000 lives, as compared with 5000 deaths from hurricanes. In 1974, however, only 27 persons were killed by tornadoes, the lowest number since the National Weather Service started keeping count, in 1922. The average number killed annually in the past five years has declined to 91, a drop of 27% from the preceding 5 years. The decline is attributed to research that allows better warnings, not to a decline in the number of tornadoes.

In fact, the number of tornadoes reported in the country has been rising steadily in recent years, mainly because of better detection of the storms. In 1974, 744 tornadoes were reported, 100 more than in the decade earlier, and nearly 500 more than in 1952. Meteorologists now think that if the full number were known, the count might be as high as 900–1000 twisters per year.

They occur in any month, but more than one-half of them come in April, May, and June, when warm air surging up from the Gulf of Mexico is most likely to clash with cold air still pouring down from the North. They form mainly where these contrasting air masses meet, usually in the Midwest and South, although tornadoes have struck in every state, a particularly vicious one hitting in suburban Washington, D.C.

Tornado forecasters still have a long way to go. They would like to differentiate between big and small tornadoes. About 350 tornadoes per year are insignificant, as small as 3–4 ft wide and lasting only 5 min. A tornado forecast now encompasses an area 140 miles wide and 180 miles long, over 25,000 square miles in all.

At the National Severe Storms Laboratory in Norman, Oklahoma, scientists are adapting a special type of radar for tornado detection. Conventional weather radar cannot "see" tornadoes, only thunderstorms. The new equipment, known as Doppler radar, will be able to determine the directions of tornadic winds by looking at the motion of raindrops in the storm. Counterclockwise winds, indicating a tornado, will show up as a unique pattern on the radarscope.

A new severe-storm indicator is wind shear—the relative directions and speeds of lower- and upper-level winds. If on the ground there is a south wind, but at 30,000 ft it is blowing from the West, there is a built-in twist in the atmosphere that can lead to tornadoes. One reason the East does not have so many tornadoes is the absence of wind shear. In the Midwest and South, wind shear is produced by the Rocky Mountains, which block the normal west–east flow of air at low levels. Surface winds in those regions often blow from southerly directions, but the high-altitude winds that clear the mountains usually blow from the West.

A tornado warning device is being developed at the Environmental Research Laboratories in Boulder, Colorado. It is a radio receiver designed to pick up a special signal broadcast by some thunderstorms. The signal's presence indicates that a thunderstorm is likely to produce tornadoes. The new detector can warn against 73% of the tornadoes that occur within a 40-mile radius of the instruments.

4-6 WINDS

"Monsoon" comes from the Arabic word for season, "mausim." The Arabs gave the name to the seasonal winds of the Arabian Sea, which blow for approximately six months from the Northeast and six months from the Southwest. In India, the word refers both to the winds and to a season of rains, and, while there is also a winter monsoon, when Indians speak of the *mausem* they generally mean the drenching rains that come with the moisture-bearing southwest winds of summer.

India's most spectacular weather event—the torrential rains of the summer monsoon—comes about because of an extreme temperature difference in the air over land and sea. When the hot air rises and leaves a vast low-pressure zone below it, enormous amounts of warm, wet air rush into the breach from the Indian Ocean and the Bay of Bengal. These winds continue, blowing northeast across the subcontinent. Along the way, their clouds may dump water on some places and pass over others, but the real downpours occur when the winds encounter the high Himalayas in northern India and, as a consequence, rise to cooler altitudes, where their vapor condenses into water. Much of the wind is then deflected toward the Northwest and travels along the Ganges and the barrier formed by the mountains.

Few studies have been done to explain why rainfall is abundant during some summer monsoons and not during others. The winter monsoon is simply rain in winter. The summer monsoon is preceded by several months of hot weather. From the end of February, the sun starts getting hotter and spring gives way to summer. The sun goes on, day after day, from east to west, scorching relentlessly. The earth cracks and deep fissures open their gaping mouths asking for water, but there is no water—only the shimmering haze at noon making mirage lakes. The sun makes an ally of the breeze, heating the air till it becomes the "loo"—the hot wind that blows between April and June and precedes the monsoon. The "loo" drops; the air becomes still.

Then comes the storm itself. In furious sweeps, it smacks open doors and windows, banging them forward and backward, smashing their glass panes. Thatched roofs and corrugated iron sheets are borne aloft. Trees are torn up by the roots and fall across power lines. This all happens in a few seconds.

Once the monsoon has arrived it stays for 3–4 months bringing frequent rains. The earth becomes a big stretch of swamp and mud. Wells and lakes fill up and burst their bounds. In the towns, gutters get clogged and streets become turbid streams. In the villages, the mud walls of huts melt in the water and thatched roofs sag. Rivers, which keep rising steadily from the time the summer's heat starts melting the snows, suddenly turn to floods as the monsoon spends itself on the mountains. Roads, railway tracks, and bridges go underwater. Houses near the river banks are swept down to the sea.

Then, almost overnight, grass begins to grow and leafless trees turn green.

While the monsoon lasts, the showers start and stop without warning. The clouds fly across, dropping their rain on the plains as it pleases them, until they reach the Himalayas. They climb up the mountainsides. Then the cold squeezes the last drops out of them. Lightning and thunder never cease. In late August or early September the season of the rains gives way to autumn.

Seventy percent of India's rainfall comes with the monsoon season, which lasts roughly from mid-June to September. Nothing short of living through it can fully convey all that it means to people for whom it is not only the source of life, but also their most exciting impact with nature. What the four seasons of the year mean to the European, the one season of the monsoon means to the Indian.

For three years (1970, 1971, 1972) the monsoon failed. Even in the best of years the rain may be excessive in some places, like Chirapunje in Assam, which gets 428 inches per year, while parts of Rajasthan, Gujarat, and Maharashtra may not get a drop. But rainfall in those three years was far below normal in most areas of the country, and buffer stocks of wheat built from surplus raised in the northern states of Punjab and Haryana, along with imports from the United States, began to dwindle. In 1972, the harvest from the "Green Revolution" belt in the North also fell below expectations. Yet in some years the rain comes in excess, even bringing floods to the center of Bombay. India lacks flood control works, or water storage in extremely large amounts, and might gain much from a study of the lessons learned earlier through the activities of organizations such as the Tennessee Valley Authority, the Bureau of Reclamation, and the Missouri–Mississippi Valley Authorities.

Monsoons are prevalent in Indochina and the countries adjacent to India. Climate modification is still one of our unfulfilled dreams, but we do know how to collect, store, and distribute the water so fundamental to our existence.

The meteorologists' term is "intertropical convergence zone," where the trade winds from the north and south meet in the vicinity of the equator, causing updrafts and clouds. When the zone itself drifts northward over Africa, the rainy season begins along the southern border of the Sahara. From June to October the parched lands bloom and the nomads' cattle and goats grow fat.

When the air currents of the intertropical convergence zone lose some of their strength, or when the water temperature of the nearby Atlantic drops slightly, the moisture content of the low-level winds is catastrophically lowered as they feed into the intertropical convergence zone. A weakening of the much higher and faster easterly winds that blow over the continent affects Africa's rains and India's monsoons.

A small, and perhaps semipermanent, perturbation in climate has been catastrophic for Mali, Senegal, Mauritania, Niger, Upper Volta, and Chad: the rains have failed. Of the nearly 3 million head of cattle in Upper Volta, only an estimated 500,000 remain; the human population of Mauritania's capital, Nouakchott, is 120,000, swollen to three times its normal size as nomads drift in from the hinterlands to wait stoically for handouts; in parts of the region the desert itself is moving south at a rate of 30 miles per year.

In recent years, immunization campaigns and newly dug water holes permitted the herds to increase. The resulting overgrazing was then compounded by the failure of the rains. Forage began to disappear, water holes to dry up, livestock to die. The nomads moved their dwindling herds southward. More than 35% of Mauritania's cattle died; some 40% have been moved to Mali and Senegal.

Can these tribesmen find any land in all of West Africa which could possibly support them? The answer seems to be "no". They flock to the towns and cities and to the rivers, seeking water, food, and medical help.

We have had "dust bowls" in Oklahoma, now the land of green fields, lakes, rivers, and canals made by man's intervention and technology. The dust bowls and the stricken Okies have been forgotten in the United States.

Meteorologically, a high-pressure area slides in from the Pacific or Canada to stall in the Nevada–Idaho–Utah plateau; simultaneously, the atmospheric pressure is lower off the Southern California coast. In search for balance, air shifts from high to low pressure, thereby creating the northeasterly wind through the basin that stretches from Santa Barbara to Mexico.

In 1970, the wind rose to 72 mph in parts of the Los Angeles basin; humidity dropped to 2% and the temperature topped 100°F. The first fire was spotted by a motorist who saw flames spreading from trash dropped near a Malibu station. Gusts in the canyons north of Los Angeles jumped sparks around like flashes from a welder's torch. Shingles torn from burning roofs flew 2500 ft high before landing a mile or two away and starting new fires.

Eight days went by before the winds abated. The toll was: 14 persons dead in Los Angeles County, and 300 firemen, one of every three on the lines, had suffered injury. A total of 795 homes and buildings had been leveled. For some it was the third time they had seen their homes besieged by flames. Property damage was nearly $100 million. Nearly 500,000 acres had been denuded into endless hills.

From five to ten times per year in the Los Angeles basin there is freedom from smog, temperatures reach 100°F, the mountainside brush is dry, and the usual sea breeze is replaced by winds from the northeast. The Santa Ana winds of California are named after a mountain pass they sometimes roar through. They have flattened hundreds of oil derricks, wrecked harbor boats, and sent glider pilots to 47,000 ft.

The same type of wind is found in many places. Oregon has its east wind. On the eastern side of the Rockies, in Wyoming and Montana, it is the "chinook". The vent d'Espagne torments southern France, the Canterbury northwester plagues New Zealand, the *zonda* works its sorcery in Argentina. For these winds the scientific name is foehn. It denotes a moving air mass which, after crossing mountains, regenerates itself into a hot, dry, gusty wind.

A foehn can last for a few hours, or up to two weeks. On one visit it hugs the mountain slopes. On the next it sails out at 75 mph past the crest before dive-bombing the earth. It once hammered Innsbruck, Austria, with 80-mph blasts that derailed street cars. Its velocity spirals so unexpectedly that monitoring instruments have broken down trying to keep up with it. It may sweep dirt and debris up in zero-visibility "walls" hundreds of feet high, while only 1 mile on either side the air is soft and clear.

Firemen are particularly weather sensitive because they know about the infernos that often follow the wind. The Swiss town of Glarus was once destroyed by a fire storm racing on the heels of the foehn. In the United States, the east wind swept the great Tillamook fire over 311,000 acres of Oregon forest in 1933. The Los Angeles basin has had Santa Ana-linked fires in 60 different years since 1900.

Desert sandstorms, in which a wall of windblown sand marches across the desert, blotting out the sun and choking man and beast, are known as "haboobs" (Arabic for "violent wind"). They are particularly well known in the area around Khartoum in the northern Sudan. It now seems that there are storms in the southwestern United States that should also be called haboobs.

A haboob is announced by a sudden change in the speed and direction of the wind. The air grows humid, the temperature drops, and what looks like a solid wall of dust roars in over the countryside. The dust is raised from the arid land and driven along at as much as 45 mph by an outflow of rain-cooled air from a towering cumulonimbus cloud.

In the United States, haboobs are usually caused by downdrafts from thunderstorms that originate in the Sierra Madre Occidental range in northern Mexico and move northwest, forming a squall line near Tucson, Arizona. The line of storm cells, each with its own downdraft and churned-up dust cloud, moves northwest, the individual clouds merging to form a dust front that can extend as high as 8000 ft. Within the dust cloud the visibility is reduced to a few hundred yards.

On July 16, 1971, a haboob formed southeast of Tucson, passed through that city at 3:30 P.M., and moved into the valley of the Santa Cruz River, which flows northwest to Phoenix. Along the path its mean speed of advance was 29 mph. It reached Phoenix at about 7:00 P.M. The relative humidity jumped from 33% to 74% and the temperature dropped more than 23°F in only 7 min. The storm's leading edge was composed of at least three "macrolobes": distinct, arc-shaped fronts of cold air churning up surface soil and dust; each of these was composed, in turn, of smaller "microlobes." Aircraft put the maximum height of the dust cloud at 8000 ft; radar indicated that some of the storm cells generating it reached 55,000 ft.

THE INTERFACES

5-1 ISLANDS

Up to this point, the earth's crust has been considered from the dynamic viewpoint. From the static viewpoint, humans have found islands—bodies of land surrounded by water—quite hospitable. The islands of the world range from minicontinents, such as Madagascar and New Guinea, to microcosms, such as Hong Kong. The major islands of the crust are listed in Table 34.

In the strict sense of the word, all continents are islands. They differ from other islands only in size. The largest islands in the world are, in square miles: Greenland (840,000), New Guinea (317,000), Borneo (287,400), Madagascar (230,035), Baffin (183,810), Sumatra (182,860), Honshu (88,000), and Great Britain (84,160).

Islands in minor waters (in square miles) are: Manitoulin (Canada): Lake Huron (1068); Singapore: Singapore Strait (224); Isle Royale (U.S.): Lake Superior (209.9); Penang (Malaysia): Strait of Malacca (110); Staten Island (64); and Manhattan (31). Atolls are coral-formed islands surrounding lagoons. The following are typical examples (in square miles): Bikini (U.S.): lagoon (280), land (2.87); Canton (U.S., U.K.): lagoon (20), land (4.3); Christmas (U.S., U.K.): lagoon (89), land (184).

5-2 BEACHES

In 1968, Congress directed the US Army Corps of Engineers to conduct the National Shoreline Study. When the Corps delivered its report in 1971, we learned that 20,500 miles of national and territorial shores are experiencing significant erosion. That is one-fourth of all our shoreline, and, excluding Alaska's 15,400 miles, the eroding portion is up to 43%.

Table 34. Major Islands and Archipelagos of the World
(*Courtesy, National Geographic Society*)

Ocean or Sea	Country	Island	Area (square miles)
Arctic	Canada	Baffin	183,810
		Ellesmere	82,119
		Victoria	81,930
	USSR	Novaya Zemlya (archipelago)	31,500
	Norway	Svalbard (archipelago)	24,100
Atlantic	United Kingdom	Great Britain	84,186
	Ireland and	Hebrides (archipelago)	2,662
	United Kingdom	Ireland	32,598
	Brazil	Marajo	1,553
	Canada	Anticosti	3,043
		Cape Breton	3,970
		Newfoundland	43,359
		Prince Edward	2,184
	Chile and Argentina	Tierra del Fuego	18,800
	Denmark	Greenland	840,000
	United Kingdom	Bahamas (archipelago)	4,400
		Falklands	4,618
		South Georgia	1,470
	Portugal	Cape Verde Islands	1,557
	Spain	Canary Islands	2,808
	Iceland	Iceland	39,768
	United States	Long Island, N.Y.	1,723
Baltic Sea	Sweden	Gotland	1,225
Caribbean	United States	Puerto Rico	3,435
	Cuba	Cuba	43,038
		Isle of Pines	1,180
	Haiti–Dominican Republic	Hispaniola	29,530
	Jamaica	Jamaica	4,232
	Trinidad	Trinidad	1,864
Indian	India	Andamans	2,508
	Ceylon	Ceylon	25,332
	Malagasy Republic	Madagascar	230,035
Mediterranean	France	Corsica	3,367
	Greece	Crete	3,207
		Euboea	1,456
	Italy	Sardinia	9,194
		Sicily	9,817

Table 34 *(continued)*

Ocean or Sea	Country	Island	Area (square miles)
	Spain	Balearics	1,936
	Cyprus	Cyprus	3,572
Pacific	Ecuador	Galapagos	2,868
	China	Hainan	13,000
	France	New Caledonia	7,335
	Canada	Vancouver	12,408
	Fiji	Viti Levu	4,053
	United Kingdom	Guadalcanal	1,130
		New Hebrides	5,700
		(United Kingdom and France)	
	Taiwan	Formosa	11,885
	United States	Aleutians	6,821
		Hawaiian	6,439
		Hawaii	4,030
		Kodiak	5,363
	Samoa	Samoan Islands	1,173
		Western Samoa	1,097
Japan Sea	Japan	Japan	142,727
		Hokkaido	30,077
		Honshu	88,000
		Kyushu	13,768
	USSR	Sakhalin	28,597
Philippine Sea	Philippines	Luzon	46,636
		Mindanao	39,191
		Negros	5,041
		Palawan	5,693
		Samar	5,309
South Pacific	Australia–Indonesia	New Guinea	317,000
	Australia	Tasmania	26,363
		New Britain	14,600
	Indonesia	Celebes	72,987
		Java	48,763
		Moluccas	28,767
		Sumatra	182,860
	Indonesia, Malaysia, and United Kingdom	Borneo	287,400
	New Zealand	North	44,281
		South	58,093

Geologists and engineers have concluded that the only sensible recourse for overdeveloped and eroding beaches is to keep pumping and dumping sand to feed the littoral drift. The sand for beach nourishment would come from the dredging of harbors, inlet shoals, and offshore sources such as bars, shoals, or the ocean bottom itself.

Hurricanes and great storms may wash sand over a barrier beach into coastal lagoons to form shoals. These shoals will eventually become part of the barrier's backside, bringing the beach system back into equilibrium. Strong winds may blow sand from the beaches into inland dune areas. Coastal erosion may eventually move the shoreline landward until the dunes again become part of the beach. Strong storm-wave action may set up powerful return currents that wash sediment away from the beaches and out onto distant shoals or the continental shelf. Some coasts, like California's, have submarine canyons close to the shore which trap beach sand. Geological analysis of Miami Beach sand has shown that it came, ages ago, from the Piedmont regions of the Carolinas.

During the depression, the Civilian Conservation Corps built a line of high dunes along the Outer Banks of North Carolina to stabilize the barrier islands. When Cape Hatteras National Seashore was established in 1953, the National Park Service built the dunes even higher. The raised dunes were soon discovered to be causing erosion. By preventing the overwash of sand, which would naturally build up the backside of the islands to replace what was lost to storms on the seaward side, the actual erosion of the beach had been intensified because the higher dunes steepened the angle of wave attack.

On the West Coast, where rivers and streams flow directly into the sea and should be the major source of beach sediment, flood-control and water-supply dams are estimated to trap 50% of the sediment that once streamed out onto the beaches. The rapid urbanization and paving of the California coast is also contributing to beach erosion by reducing the production of natural debris that would have eroded into the sea. Inman and Frautschy[1] illustrated five littoral cells along the California coast (Figure 18). Southern California's weather runs in wet-and-dry cycles of about 25 years, corresponding roughly to the double sunspot cycle that effects earth's weather. In a dry cycle, little sediment gets washed into the sea.

Drought, hurricanes, and tsunamis (seismic sea waves generated by earthquakes) can periodically pose serious problems. Sea-level rise and crustal subsidence may point to a slow encroachment of the sea over the very long run. All of the low coastal area is more or less in equilibrium with hurricane conditions. The Army Corps of Engineers has observed that erecting a building on a beach is like building on an active volcano. Sooner or later, the structure is lost.

[1]D. B. INMAN and J. D. FRAUTSCHY, *Santa Barbara Coastal Engineering Conference*, A.S.C.E., New York, 1966, pp. 511–536.

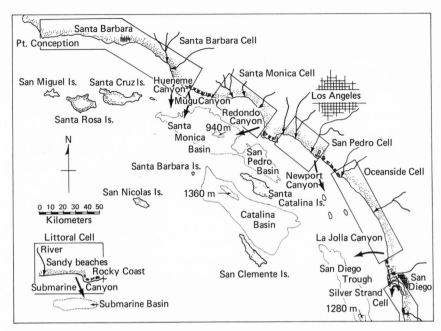

FIGURE 18. Five littoral cells along the Southern California coast. Each cell contains a sedimentation cycle. Most sand is brought to the coast by streams, carried along the shore by waves and currents, and lost into offshore basins by submarine canyons.

Natural forces have never made a permanent seashore or lake beach and apparently never will. A beach is destined to move, change its shape, thinning out in one spot and gradually reappearing elsewhere. This continuously changes our seashore maps.

The Army Corps of Engineers thinks of the very popular beaches of the southern coast of Long Island as one beach. Put a jetty to hold sand at one place, and the sand builds up on the updrift side and moves from the downdrift side. The beaches do not know of political boundaries.

Shinnecock Bay is the first of a group of shallow bodies of water that run behind the beaches 35 miles west of Montauk Point, the easternmost point on Long Island. There are beaches all the way from Montauk to Fire Island: Jones Beach, Long Beach, the Rockaways, and Coney Island at the southeast corner of New York Bay. The Jersey beach, formed at right angles to Long Island, runs from Sandy Hook along the Jersey coast to Long Branch, Asbury Park, Atlantic City, and on to Cape May, the southernmost point.

From the air at Montauk, Long Island, you can see the outer sandbar, parallel to the beach. Whales were caught commercially at Amagansett until 1918. The sea has been eroding the land at Montauk Point for years, moving the ends of the island westward and depositing sand where the currents are interrupted.

At Moriches, one sees deltas of sand deposited by currents. They are inside the breachways, driven there by the sea. They are in protected backwaters. Shinnecock had such a tidal delta, but the chief hydraulic phenomenon was a large, rough sprawl of sand fanning out from the inlet, pushing into the teeth of the Atlantic fetch and carving out an ever-greater inlet there.

The sea has been eroding the land on Long Island for years, moving the ends of the island westward and depositing the soil where its current is interrupted. It used to have a free run along two-thirds of the length of the island before it was intercepted by the build up of sand off Fire Island Inlet. Since the 1938 hurricane, however, there has been a natural breachway at Shinnecock. The current pushing out and the jetties have caused the river-like littoral current to drop much of its load of sand earlier than it used to.

The prevailing wind is southwest; the shore tends to move from New York harbor east toward Montauk. This ignores the heavier "nor'easters" and their unlimited fetch in from the Aran Islands. The warmer and therefore lighter "sou'westers" have their fetch mitigated by the Jersey coast.

Flying along the south shore toward Jamaica Bay, these inlets are repeated a half dozen times. There are some 4 miles of sand between Fire Island Light and the inlet at the edge of which it was built in 1857. Although the littoral current is reversed from west to east, the same thing is happening along the New Jersey shore at Barnegat, Sandy Hook, and Cape May.

The sea not only moves the sand laterally, it also moves it in and out. Much of the sand spent by the littoral current along the beach comes from the withdrawing phase of this in-and-out process. Late in the year the winds are cold, making the waves heavy and steep so that they hit deeper into the beach in places where the summer waves have not reached, to flatten and pack the sand. This soft, vulnerable pile lies just before the dunes. The same process happens along the entire seacoast of the United States. Beaches move, grow, retreat, and rebuild. They might be considered as moving crust, not yet converted to sedimentary rock.

Gofseyeff and Johnson[2] developed a map showing that longshore sand transport has caused an extension of the Fire Island spit at a rate of 100 m per year since 1825, as shown in Figure 19.

Inman and Brush,[3] of the Scripps Institute of Oceanography, have discussed the coastal zone, the interface of the hydrosphere, the lithosphere, and the atmosphere, the physical processes in nearshore waters, the interchange of energy in wave forms, sea currents, and the transport of sediments and sands,

[2]S. GOFSEYEFF and J. W. JOHNSON, *Proceedings of the 3rd Conference on Coastal Engineering*, Council on Wave Research, University of California, Berkeley, 1953, pp. 272–305.
[3]D. L. INMAN and B. M. BRUSH, *Science*, *181*:20–34 (1973).

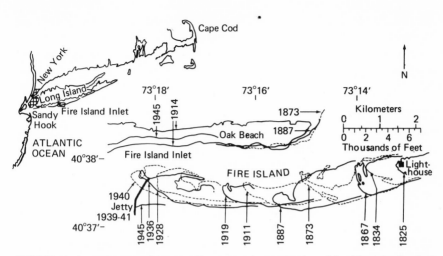

FIGURE 19. Longshore sand transport has caused an extension of the Fire Island spit at an average rate of 100 m per year since 1825. (Courtesy, J. W. Johnson, University of California, Berkeley)

and circulation over shelves. Figure 20 gives coastal zone nomenclature, Figure 21 an energy balance in a coastal zone, and Figure 22 a circulation cell.

Although the continental shelves and nearshore waters comprise only about 5% of the area of the world, about two-thirds of the world's population lives near the coast.

The coastlines of the world look craggy and indomitable, the rocky stretches seem tough, and the beaches appear to be permanent. But the sand on the beaches is easily moved and is in increasingly short supply. In reality, the beaches are fragile ribbons of sand that are frequently broken by acts of nature.

The total rate of dissipation of mechanical energy in the shallow waters of the world is about 5×10^9 kW (Figure 21). Dissipation occurs over the entire shelf. Surface waves dissipate their energy primarily nearshore, whereas tidal and other ocean currents and internal waves dissipate most of their energy over the outer portions of the shelf. Internal waves, edge waves, shelf seiches, and local winds may produce water motion over the continental shelf and submarine canyons. A seiche is an oscillation in a body of water for a period of minutes or hours, depending on the given area and depth of the basin. Seiches may disturb shipping or moored structures or may drown unwarned persons on piers or shores.

The principal sources of beach and nearshore sediments are: rivers; unconsolidated material from sea cliffs, which are eroded by waves; and shells, coral fragments, and skeletons of marine organisms. Many beaches, such as those along the east coast of the United States, are supplied by sand that has been

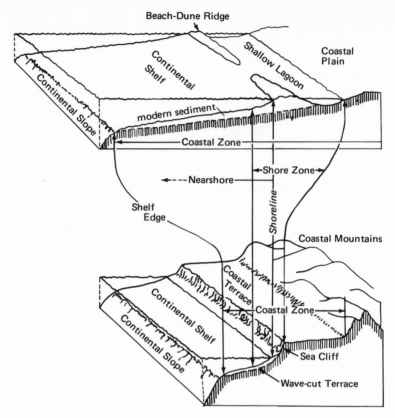

FIGURE 20. Coastal Zone nomenclature. Wide-shelf plains coast (upper part) and narrow-shelf mountainous coasts (lower part) are characteristic of the East Coast trailing edge and the West Coast collision edge. (Courtesy, Inman and Nordstrom)

reworked by waves and currents from ancient river and glacial material deposited during former stillstands of sea level.[4]

Streams and rivers are by far the most important source of sand for beaches in temperate latitudes. Cliff erosion probably does not account for more than about 5% of the material on most beaches, except locally on trailing-edge coasts such as the east coast of the United States.

In Back Bay, just south of Virginia Beach, Virginia, fishing and crabbing were once worthwhile commercial enterprises. Ducks and geese made Back Bay a regular stop in their migrations because a large part of it is a national game preserve. Slowly the sandbars and channels at the mouth of the bay

[4]C. P. IDYLL, *Scientific American*, 228:6 (1973).

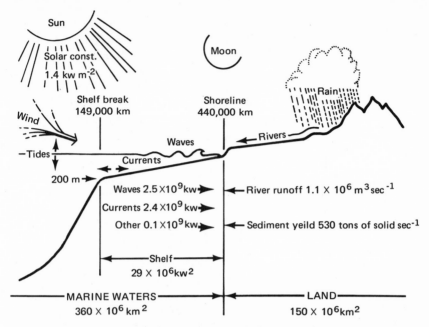

FIGURE 21. Energy balance and land runoff in the coastal zone. Most of the energy is supplied by the open sea.

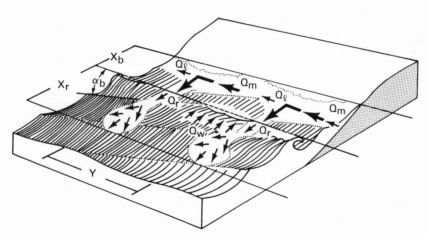

FIGURE 22. Model when waves break at angle to the beach. The zone between rip currents becomes a circulation cell. Primary mixing occurs in the surf ($x_b \cdot$ y); secondary mixing at ($X_r \cdot$ Y).

shifted as a result of hurricanes. The shifting reduced the tidal effect in the bay, thus lowering the salt content of the bay's water. The fishing and crabbing stopped being profitable. The ducks and geese stopped coming. The city of Virginia Beach decided to save the bay by pumping ocean water into the head of the bay, thereby raising the salt content. The procedure worked. Through constant monitoring of salt levels and continuous pumping, Back Bay has never been in such good condition. The commercial fishing has resumed, and the ducks and geese have come back in greater numbers than ever before.

5-3 EL NIÑO

The El Niño wind, or cycle, is an example of interfacial effects at the junction of land, atmosphere, and sea. There is said to be a 7-year cycle but the changes are less precise. It may be repeated for 2 or more years or may not occur in a decade. Changes usually occur at Christmastime; El Niño means "the Christ Child."

The wind now comes from the West rather than from the Southeast, and it is laden with moisture from the Pacific. With no mountains to rob the air of its burden of water, the arid coast is subjected to torrential rains and severe windstorms. In a desert region, where even a heavy mist can cause problems, the floods of El Niño are often devastating. In some Niño years no rain falls with the wind change.

In the period from 1961 to 1971, the world's largest fishery had been in the Humboldt current off the coast of Peru. Fish meal has been important for chicken raising in the United States and Europe as well as for animal foods. The world of the Peruvian anchovy is the sweep of a cold northward drift, in company with tiny plants and animals on which the fish feed.

The Peru coast is kept cool by the 10°C (50°F) ocean current which reaches 22°C (71.6°F) in the North, as shown in Figure 23. The average air temperature is 18–22°C (64–72°F). Along the nearly 1500 miles of coastline, there are no marshes, mudflats, or estuaries, but only a treeless, barren brown desert of sand, so fine that the natives can ski on it. The desert rarely extends more than 40 miles to the Andes, whose high rain shadows catch the prevailing southeast winds and take their moisture. A few small streams and their narrow valleys are green, but they are surrounded by sand.

The four components of the Peru Current are shown in Figures 23 and 24. The coastal current runs from about Valpariso, Chile, in the south, to north of Chimbote, Peru, or some 2000 miles. The anchovies live mostly in the northern part of this band of water, which constantly changes shape and size. The Oceanic current is longer than the Coastal current. It is often several

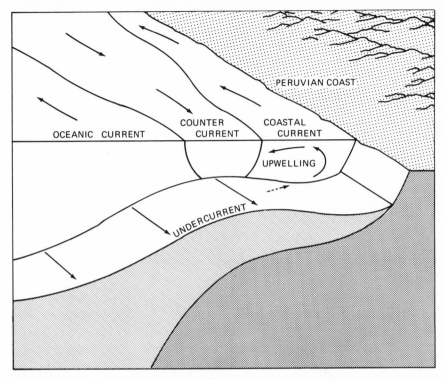

FIGURE 23. Two south-flowing and two north-flowing components of the Peru Current. The countercurrent is a surface or near surface stream of water, intruding between the north-flowing Coastal and Oceanic current and extending not much south of the equator. When the wind that moves the north currents falters or changes direction, the warm water pushes far to the south. Deep beneath all three currents is the second, larger south-flowing undercurrent.

hundred miles wide, and it runs as deep as 700 m. It flows north to the Gulf of Guayaquil before bending west. Much of the cold is the cold of subsurface water. As the water on the ocean surface is swept away by the prevailing winds, deeper low-temperature water slowly wells up to replace it. The trade winds blow from the South and Southeast, mostly parallel to the shore. This prevailing wind urges the surface water northward. Offshore drift skims off the surface layer and the cold subsurface water rises to replace it, traveling vertically at a rate ranging from 20 to 100 m per month.

The biological effect of the upwelling is enormous. That stretch of water produces fully 22% of all the fish caught throughout the world. Phosphates and nitrates stimulate plant growth. Nutrients, accumulated gradually in the deep layers of the ocean as the debris of dead marine plants and animals sinks to the bottom, travel with the upwelling water to the top levels. There, the light is sufficient for photosynthesis, and the nutrients help the marine plants to

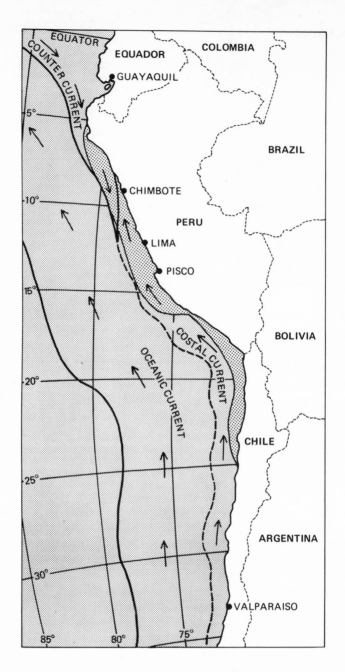

FIGURE 24. Two north-flowing components of the Peru Current are the deep, narrow Coastal Current that hugs the land from Valparaiso, Chile, to north of Chimbote, Peru, and the deeper and wider Oceanic Current that reaches the latitude of the Gulf of Guayaquil. The anchovies are normally found in the Coastal Current between 25° and 5° South latitude, and may consist of a northern and a southern population.

flourish. The concentration of nutrients in the Peru upwelling is many times greater than that in the open ocean. The amount of carbon per cubic meter of water per day is fixed at 45–200 mg, as compared with less than 15 mg in the adjacent waters. Perhaps only the Benguela Current off the southwestern coast of Africa is richer.

The Peru upwelling sustains an enormous flow of living matter. The food chain begins with the microscopic diatoms and other members of the phytoplankton that comprise the pasturage of the sea. The plants grow in profusion. The food chain can go on and progress, from the small fishes that eat the herbivores to the larger fishes and squids that prey on the small fishes. Most of its energy flow stops with the anchovies. This species has captured a high proportion of the energy available. At the height of the anchovies' annual cycle, the bulk of the catch is probably of the order of 15–20 million metric tons.

The anchovy schools do not move at random; apparently because of a strong preference for the cold water of the Coastal current they remain within a comparatively restricted zone. The Coastal current is at its narrowest during the southern summer, running close to shore and seldom exceeding 200 m in depth. The anchovies press together in enormous concentrations near the shore and close to the surface. The larger fishes and the squids feed well; the guano birds have to fly only short distances from their island nesting grounds and need not dive deep to reach their prey. The fisherman finds summer work the easiest, often setting a purse seine within sight of port and gathering in anchovies 100 tons at a time.

A warning of sea change is given when the temperature of the coastal water rises. The world of the anchovy tilts. With the warmer water come unfamiliar inhabitants of the northern Tropics: the yellowfin tuna, the dolphin-fish, the manta ray, and the hammerhead shark. Some feed on the anchovies. A greater threat is a slowing of the northbound Coastal current and a decline or even a halt in the usual upwelling of subsurface waters. As the supply of nutrients diminishes, the planktonic plant life that provides the base of the ocean food chain becomes less abundant. Furthermore, the water temperature is now too high to suit the anchovies. Even if the shortage of food has not yet greatly reduced their number, the fish scatter.

Sea changes result when the steady southeast trade winds weaken or when the wind blows from the West. The ocean currents to the northwest are no longer pushed along with the same vigor; the south-flowing Peru countercurrent is weak, but when the prevailing winds fail or are reversed, the countercurrent thrusts a tongue of warm water into the cleft between the now less vigorous Coastal and Oceanic currents. As it meets less resistance, the countercurrent penetrates farther south, pushing the weak north-flowing currents aside and covering their cold waters with a 30-meter layer of warm tropical water.

The anchovy industry began in earnest in 1957. Within 10 years the profits that could be made from catching and processing the fish attracted hundreds of fishing boats and led to the construction of dozens of fish-meal factories.

After the Niño year of 1965, the fishery had enjoyed several successful seasons, culminating, in 1970, with an anchovy catch of 12.3 million metric tons. Then, toward the end of April, 1972, fishing suddenly faltered. By the end of June catches had dwindled to almost nothing, and at the close of the 1972 season only 4.5 million tons of anchovies had been harvested.

Quite apart from the sea change, however, the Peruvian commercial fishery must accept a share of the blame. The 1970 catch of 12.3 million tons considerably exceeded the 10-million-ton level that fishery biologists had estimated to be the maximum sustainable yield of the Peruvian stock. Losses from spoilage at sea and unloading and processing ashore raise the commercial total to some 13 or 14 million tons. There are many more fishing boats and fish-meal factories in Peru than are needed to harvest and process the catch. The anchovy fleet is so large that it could harvest the equivalent of the annual United States catch of yellowfin tuna in a single day, or the annual United States salmon catch in two-and-a-half days. The fleet could be reduced by more than 25% and still comfortably harvest a rational quota of 10 million tons of anchovies per year.

5-4 SWAMPS AND MARSHES

The two largest swamps in North America are the Okefenokee and the Atchafalaya. But, partly because of a comic-strip opossum named Pogo, the Okefenokee of Georgia is vastly better known.

And yet the Atchafalaya Basin swamplands and coastal marshes cover almost twice the area of the Okefenokee. Through the heart of the basin flows the Atchafalaya River, carrying one-third of the water draining off almost half of the coterminous states, a torrent third only to the Mississippi and the Columbia among American rivers. Until its connection with the Mississippi was dammed, the Atchafalaya was well on its way toward becoming part of the Mississippi itself instead of one of its distributaries.

From the juncture of the Mississippi and the Atchafalaya, near Natchez, the route to the Gulf by the distributary is only 145 miles, 177 miles shorter than by the main stream. If the Corps of Engineers had not stopped the drift with a control dam, by 1970 the Atchafalaya would have captured the Mississippi and become the main stream, leaving Baton Rouge and New Orleans stranded as inland cities on a sluggish bayou.

The Corps of Engineers did not entirely cut off the Atchafalaya from the Mississippi. The basin must be available to soak up the waters of Project Flood. Engineers define Project Flood as the worst possible hypothetical flood any combination of upriver rains and thaws could pour down the Mississippi. To save Baton Rouge and New Orleans, the gates would open to let through 1.5 million cubic feet of water per second, a torrent greater than the main river itself could handle at New Orleans. That immense flood would drown the entire Atchafalaya Basin.

Even normal high waters of a typical spring overflow the river's banks and flood the basin with water from 6 inches to several feet deep. The phenomenon simultaneously gives birth to the rich swamp ecosystem—and, paradoxically, kills it. For the swamp is dying. Once it reached north into southeastern Arkansas, and already it has dried to one-fourth of its former size. Since World War II, soybean fields have spread over areas of the northern end.

Even the lowest spring flood carries a tremendous load of silt stripped from the whole Mississippi and Red river systems. When flood currents slow, that silt settles and raises the land level, inexorably and rather rapidly changing the swamp to upland woods and, eventually to farmland. The Corps of Engineers must prevent that buildup. The Atchafalaya must remain low so as to absorb Project Flood, if it should come.

Dikes have been proposed to contain the Atchafalaya's spring torrents, with openings every 2000 ft or so to permit controlled escape. Flanking dikes would turn the escaping water sharply back upstream, forcing a rapid deceleration and deposit of silt right at the escape point, where it could easily be dredged and added to the dike as spoil.

The enemy here is not industrial use but the laws of geological change. A swamp is not a static phenomenon, but, like every natural feature on this planet, forever changing. A swamp simply changes faster than does a mountain, hence the change is more noticeable. Like all swamps at the delta of a river, the Atchafalaya, as it now exists, is doomed. Inexorably, the river drops its silt, building land on both banks until it has contained itself within natural levees.

Elsewhere on the Louisiana coast the sea erodes valuable estuarine marshland at the rate of 16.5 square miles per year. But the silt pouring down the Atchafalaya, unwanted in the swamp, is building an immense new delta in the Gulf at the south end of the basin. It is predicted that by the year 2020 the silt will have wiped out present Atchafalaya Bay and built as much as 100–150 square miles of new land.

The same process that is now killing the basin will build a new swamp even farther southward. The mud first builds a salt marsh, then overflows build a marsh high enough to support trees, and a new swamp is born. And on the day of its birth, it begins to die—only to be replaced still farther south by a new marsh and swamp.

5-5   REEFS

Reefs are natural shelters for much of the life which makes up the ecocycle of the sea. Where nature has not provided such shelter, sea creatures are quick to adopt substitutes, be they sunken ships, old tires, cars, or the steel pilings of oil rigs.

Observation by marine scientists has shown a steady chain of sea life proliferation around such artificial reefs. Clusters of living, growing organisms attach themselves to the surfaces: barnacles, algae, anemones, and others. This food supply attracts small fish, which in turn lure the larger, carnivorous fish. The process continues until the biological community has a full representation of feeding types.

Off Louisiana, a sports fishing industry has developed around oil production platforms. The commercial fish catch in the entire Gulf has also grown appreciably in recent years. Fishermen and oilmen have found they can coexist amiably on the Gulf waters. The oil platforms are the biggest, fanciest fishing lures in the world.

Other artificial reefs are in areas lacking natural formations. The U.S. House of Representatives not long ago passed a bill to permit the sinking of up to 144 surplus Liberty ships in coastal waters. And there have been proposals to form reefs with junk cars and rubber tires. Artificial reefs hold promise for man and fish alike.

The underwater world of Florida's John Pennekamp Coral Reef State Park was founded in 1960. The park covers 155 square miles, roughly 10% of the only living coral reef in the continental United States. Molassas Park lies at the park's outer perimeter, just on the edge of the Gulf Stream. An estimated 650,000 persons visited the facility in 1972. Of these, 25,000 were scuba divers.

In March, 1960, President Eisenhower signed an order creating the Key Largo Coral Reef Preserve, the world's first underwater park. It also became known as the Pennekamp Park when the State of Florida assumed responsibility for its operation and management.

5-6   RED TIDES

"Red tides" periodically plague the west coast of Florida. In 1971, and a number of times earlier, Tampa collected 1000 tons of dead fish—some weighing 300 lb—from its beaches, and had to find burial areas for the decomposing creatures. St. Petersburg collected twice as much in its area.

Red tides occur when conditions are favorable for rapid multiplication of dinoflagellates, such as *Gymnodinium breve*. When present in great numbers they give a reddish hue to the sea. They are tiny creatures, propelling themselves with a whip-like tail. During a "bloom," sea water may contain 60 million of the organisms per liter. Tampa Bay reached 20 million per liter when lack of rain allowed Tampa Bay to become salty enough for the dinoflagellates to prosper. A high tide swept them into the bay, and Tampa suffered a red tide and stinking dead fish on its doorstep.

Shellfish can eat poisonous organisms of the red tide type and appear unaffected, but when humans eat the clams, mussels, and abalone they become paralyzed. Such poisonings occur along the entire West Coast, from Mexico to Alaska, at the time of the "red tides."

In 1971, there was an outbreak of poisoning by mussels at Yerscke, the Netherlands, and in May, 1968, along the coasts of Northumberland and Durham, England. Eighty people were affected but no one died. However, the effect on sea birds was catastrophic. Eighty percent of the breeding population of shag cormorants were killed and their bodies were everywhere.

When the red tides appear off the seacoasts, sea creatures are killed by the hundreds of thousands. These epidemic outbreaks are not well understood, but they produce more waste in one season than all the human-caused fish kills of two decades.

Red tides have been known since 500 B.C. New England waters are cold, and with a good tidal flush are normally inhospitable to flagellates. In September, 1972, red tides invaded the New England coast from Cape Cod to upper Maine and brought losses of $1 million per week. New York and Long Island beaches had a Portugese Man-of-War jellyfish invasion at about the same time and many bathers were stung. The New England states set up embargoes on the export of seafood until the red tides disappeared. More than 40 people were poisoned, but none fatally. The economic disruption amounted to many millions of dollars.

5–7 SUBSIDENCES

At Baytown, Texas, waterfront homes and much of the neighborhood have sunk 9 ft in 30 years, because of the compression of the earth as water supplies are pumped out. According to the National Oceanic and Atmospheric Administration, most sections of the country are rising or falling at a rate of less than one-half inch per year. In some areas the rate is accelerated by withdrawing huge quantities of water, natural gas, or oil, which results in dehydration and compaction of sedimentary layers of clay.

San Jose, California, has sunk 10 ft in 40 years, primarily as a result of water pumping to irrigate the Santa Clara Valley. The city is sinking uniformly and is located safely away from the sea. Long Beach, California, has sunk 27 ft in 30 years, from oil and gas pumping. The subsidence caused over $100 million in damage, buckling pipelines and railroad beds, severely damaging bridges and utility lines, until countermeasures begun in the late 1950s curtailed the movement.

Much of the Atlantic Coast is sinking a fraction of an inch annually, as is part of Florida, with little apparent effect. In contrast, the southern Appalachian area, the Rocky Mountains, and much of the Midwest are rising slightly.

The most serious current subsidence is in the Houston–Galveston area. Hundreds of homes and industrial installations have sunk to a level where they are flooded by high tides. Shell Oil Company has been forced to raise its docks along the Houston ship canal, and Exxon Corporation's giant Humble refining plant at Baytown is cutting back pumping of ground water to prevent its docks from sinking into the sea. The Army Corps of Engineers is making studies for a huge levee to close off the mouth of the bay and create a lake, thus protecting the homes from further inroads of the Gulf.

Long Beach was able to restrain the sea from threatened shipyards and other ocean-front areas with levees, and almost halted the sinking. It even gained back one-half foot in some places. This was mainly accomplished through repressuring the underground sands by pumping water back into the ground.

More than 2.5 million tons of various chemicals may be deposited on New York and New England in a year of average precipitation. Because of salt spray, sodium chloride will dominate along the coast. In rural areas inland, sulfate and calcium are much more significant, whereas near industrial areas even higher concentrations of calcium, nitrate, and sulfate are found. This is one conclusion of a survey carried out by the U.S. Geological Survey, which finds that water from rain and snow falling over relatively contaminant-free areas contains significant amounts of sea salts, sodium–calcium bicarbonates, and sulfuric and nitric acids from outside sources, and that these impurities have an effect on surface- and ground-water resources. In the Northeast, hydrogen-ion concentrations are higher in precipitation than throughout the rest of the United States.

USGS hydrologists collected samples from 18 sites in New York, Pennsylvania, and New England during periods lasting 12–36 months. The collected samples were analyzed for calcium, magnesium, sodium, potassium, ammonium, hydrogen ion, bicarbonate, sulfate, chloride, nitrate, and phosphate. Combining the analytical results with precipitation amounts, USGS scientists came up with the following annual average loadings (per day per square mile): calcium, 11 lb; magnesium, 2 lb; potassium, 3 lb; sodium, 5 lb; chloride, 9 lb; and sulfate, 76 lb.

In all cases studied the nitrogen loads in precipitation entering the northeastern basins are greater than the nitrogen loads being carried by the streams. All sulfate and nitrogen, and much of the chloride and potassium in streams—particularly those underlain by rocks that do not tend to dissolve—are supplied by precipitation.

USGS also compared precipitation with water from the Magothy aquifer, a large formation of subsurface water-bearing rocks that supplies water to the eastern two-thirds of Long Island. The chemical content of water in the aquifer is largely determined by the chemistry of atmospheric precipitation, rather than by the chemistry of the aquifer rock. The quality of the aquifer water was equal to that of rainwater.

CHEMICAL CYCLES

6 – 1   THE CARBON CYCLE AND THE BIOSPHERE

6 – 1a   Photosynthesis: Carbon Dioxide and Oxygen

Photosynthesis is the process by which plants convert the carbon dioxide in the air to organic matter, using water and the energy of sunlight. In the atmosphere, carbon dioxide is present as 3 parts in 10,000 parts of air (0.03% by volume). Through millions of small pores, or stomata, on the surfaces of leaves, air penetrates and gives up about 10% of its carbon dioxide content. The leaf cells have small particles, or chloroplasts, containing green chlorophyll, similar in structure to the hemoglobin of the blood of animals. In reflected light, chlorophyll appears red.

Carbon dioxide unites with the chlorophyll in a series of enzyme-regulated reactions, and then combines with water to form dextrose sugar. An excess of oxygen is released to the air. Sunlight provides the energy, but only 1% of the solar energy falling on the leaf is used for photosynthesis. Dextrose may be converted to other sugars, or may be combined with nitrogen to form amino acids. These are the building blocks for the proteins on which all life depends. Dextrose is also converted to starch, cellulose, fats, and other materials.

Photosynthesis is favored by temperatures about 70°F, diffused moderate light, an ample supply of water, fertile soil, and a supply of trace-mineral elements. On hot, bright summer days, the efficiency of photosynthesis goes down, and the rate also declines when the soil is dry. The reaction may be summarized as:

$$CO_2 + \text{water} + \text{sunlight energy} = \text{dextrose and oxygen}$$

Respiration is a life process of all living organisms which consists of oxidizing foods at low temperatures. In a plant or tree the food is dextrose. Thus, in respiration:

$$\text{Dextrose} + \text{oxygen} \rightarrow CO_2 + \text{water} + \text{energy}$$

In the daytime, photosynthesis and respiration occur simultaneously, with photosynthesis being more intensive. Carbon dioxide is absorbed and oxygen released. At night, photosynthesis stops but respiration continues. Oxygen is taken in and carbon dioxide is given off, as in animals. The warmer the atmosphere the more intense the respiration, so that internal plant temperatures of 120–130°F are deadly.

The CO_2, CO, and methane cycles are parts of the larger carbon cycle, expressed in the form of CO_2 in Figure 25. Some geochemists have estimated atmospheric additions and losses by the atmosphere during geologic times (Table 35).

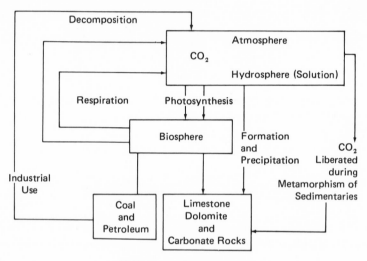

FIGURE 25. Carbon cycle

Table 35. Additions of Carbon Dioxide

Source	Amount of surface (per cm²)
Gases from volcanoes, igneous activity	0.00245 mg per year
Vital functions and decay of organisms	40 mg per year
Locked up in carbonate minerals	6562 g
Coal, bitumen, petroleum	760 g
Dissolved in hydrosphere	20 g
Atmosphere	0.4 g
Total	7350 g
Geologic time	3×10^9 years
CO_2 from volcanoes and igneous activity	0.00245 mg per year

The biosphere is that part of the earth capable of supporting life, primarily in a narrow zone at the surface of the hydrosphere (oceans, lakes, rivers) and on the lithosphere (the crust) where water, air, and radiation (sunlight) support prolific life. Life as man knows it depends on a small number of organic compounds — sugars, amino acids, purines — the building blocks of the carbohydrates, proteins, and nucleic acids, formed step-wise and at low temperatures isothermally (without temperature change). Life cycles depend on plants as producers, and on animals as predators or consumers. Plants inhale CO_2. The assimilation of CO_2 through the activity of chlorophyll and the utilization of radiant energy is of overwhelming importance for life on earth, expressed as the reaction

$$6CO_2 + 5H_2O + 678,000 \text{ cal of sunlight} = C_6H_{10}O_5 \text{ (starch)} + 6O_2$$

Mushrooms lack chlorophyll and cannot use sunlight to manufacture their food. They must feed on organic carbon compounds, such as living trees, dead logs, decaying leaves, animal excretions, and the dead needles of conifers. These materials all abound on the forest floor, making it a favorite habitat for fungi. Some fungi are poisonous to humans; however, their secrets have been learned and we now cultivate the nontoxic varieties for food.

There is geologic evidence that oxygen production has been largely the product of photosynthesis. There is an abundant record of plant fossils in sediments from early geologic periods, such as Devonian, as well as organic carbon in large quantities in Precambrian sedimentaries, among the earliest-known geologic periods.

The research vessel *Knorr,* operated by the Woods Hole Oceanographic Institution, and equipped with elaborate sampling, sensing, and recording instruments, and with computerized devices for recording water temperature, salinity, oxygen and carbon dioxide content, and sediment density, completed a 30,000-mile trip from the North Atlantic to Antarctica from July, 1972, to April, 1973. Figure 26 shows the recent voyages of the *Knorr,* on which it was found that carbon dioxide from human activity is absorbed by the North Atlantic and to some extent exhaled by the South Atlantic oceans.

There have been proponents of a "greenhouse" theory who have worried (without data) that carbon dioxide from industrial activity is building up in the atmosphere to the point that it causes serious climatic changes. Carbon dioxide is alleged to act like the glass roof of a greenhouse, permitting sunlight to pass through but inhibiting the escape of heat. But since it is known that water vapor has a similar effect, there has been little scientific acceptance of the theory.

In layers of the ocean where there is turbulence, there are associated increases in the quantities of sediment and nutrients, as in the Humboldt current off Peru. The Walvis Ridge in the South Atlantic bars the flow of nutrient-containing cold water from Antarctica to the South Atlantic. This ridge also affects the Benguela current, which is similar to the Humboldt, and the Namib

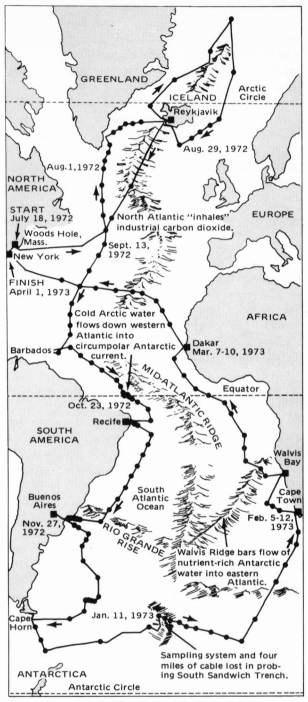

FIGURE 26. The Knorr sailed 30,000 miles on her 7-month journey, collecting samples at 121 sites.

desert of Southwest Africa. Calcium compounds and carbon dioxide are removed from the ocean waters by oysters, clams, and mussels, which form calcium carbonate shells, and by the formation of sedimentary rocks such as limestones, marble, and dolomites.

6 – 1b   Carbon Monoxide

In its latest edition, the *Merck Index,* a chemical and pharmaceutical bible of the practicing scientist, gives the properties of carbon monoxide:

> Highly poisonous, odorless, colorless, tasteless gas. Very flammable, burns in air with a bright blue flame. Melting point $-205.0°$. Boiling point, $-191.5°$. Density at $4°C^{-195}$ (liq.) 0.814. Density (gas), 0.968 (air = 1.000). d $^♭_4$ at 760 mm. = 1,250 g/liter. Cubic feet per pound at 18° = 12.8. . . . Sparingly soluble in water: 3.3 vol./100 ml. H_2O at 0°; 2.3 vol./100 ml. H_2O at 20°; freely absorbed by a concd. soln. of cuprous chloride in HCl or in NH_4OH. Appreciably soluble in some organic solvents, such as ethyl acetate. The solubility in methanol and ethanol is about 7 times as great as the solubility in water. . . . Human Toxicity: Combines with hemoglobin of the blood to form carboxyhemoglobin which is useless as an oxygen carrier. Symptoms: Headache, mental dullness, dizziness, weakness, nausea, vomiting, loss of muscular control, increased then decreased pulse and respiratory rates, erythema, collapse, unconsciousness and death. 200 p.p.m. may produce symptoms of poisoning in a few hrs. MAC [Maximum allowable concentration] for 8 hours exposure is 100 p.p.m.; 400 p.p.m. for a period not exceeding 1 hour. Concn. of 1000 p.p.m. can produce unconsciousness in approx. 1 hour and death in 4 hours. *Antidote:* Oxygen.

Nature is a much greater polluter than man, at least in terms of the quantities of carbon monoxide spewed into the atmosphere. This conclusion was reported by the Argonne National Laboratory in Argonne, Illinois, and is contrary to previously held notions that human activities are responsible for most of the carbon monoxide emitted. The researchers found that natural sources account for more than 3.5 billion tons per year of carbon monoxide, or 90% of the world's total. Human-generated quantities of the gas, in comparison, account for only about 270 million tons per year.

There are two major sources of nature's production of carbon monoxide. The largest source is by oxidation of methane, which is emitted by decaying organic matter in swamps, tropical forests, and rice paddies. This source is estimated to churn out at least 3 billion tons per year worldwide. The second-largest source is the decay of chlorophyll, which accounts for 100 million tons per year. The remainder comes from several sources, including the oceans.

There seems to be no escape from the carbon monoxide produced by methane oxidation. Worldwide geographic air samplings by the Argonne Laboratory — in the Northern Hemisphere in summer, in Australia in autumn, and in American Samoa in winter — indicate that carbon monoxide from methane oxidation is dominant the world over.

Pinpointing chlorophyll as a major source of carbon monoxide confirms research by two other Argonne scientists, who concluded that chlorophyll in algae produces large amounts of carbon monoxide. They found a large burst of natural carbon monoxide during the early autumn months in the Northern Hemisphere, when the seasonal decay of chlorophyll in green plants and tree leaves takes place.

The Argonne team found that isotopes of oxygen and carbon allowed differentiation between natural sources of carbon monoxide and those arising from our modern industrial society. The carbon monoxide produced from a gasoline engine, as it turns out, has a relatively low level of oxygen isotopes as compared with carbon monoxide from natural sources. Differences in relative amounts of isotopic carbon were also found. The identifiable species are detailed in Table 36.

Argonne researchers measured isotopic concentrations of carbon monoxide in the atmosphere to estimate how much of each kind exists in the world. The compositions were so clearly defined that analyses by mass spectrometry could easily detect the source of the gas. A basic level was established to identify the source. Urban air samples from which pollution levels could be estimated were then gathered from various points around the world, and similar samples were collected from locations where low pollution levels might be anticipated.

The Argonne group plans to use similar techniques for related studies. Isotopic measurements will be used to determine the concentration of nitrogen oxides in the stratosphere. Such a base would be necessary to determine the effect of supersonic transports on nitrogen oxide levels in the stratosphere. The group also believes that regular sampling of the atmosphere at diverse locations, especially in areas influenced by the trade winds and westerlies of each hemisphere, will uncover data that will provide a better understanding of carbon monoxide's role in nature.

Although localized buildup of carbon monoxide in cities may still represent a serious health hazard, artificial carbon monoxide production is relatively insignificant when considered on a global basis.

The Argonne project was performed under a contract from the Coordinating Research Council, a nonprofit organization directed by the American Petroleum Institute, the Automobile Manufacturers Association, and the U.S. Environmental Protection Agency.

In a similar study, a Texas A&M University team, using sonar-type instruments, identified a large number of streams of methane and ethane gas and traced them to cracks in the ocean floor about 65 miles southeast of Galveston in the Gulf of Mexico.[1-4]

[1]*Science, 177:*338–339 (1973).
[2]J. C. McConnell, M. B. McElroy, and S. C. Wofsy, *Nature, 223:*187 (1971).
[3]B. Weinstock, *Science 176:*290 (1972).
[4]R. B. Ingersoll, *Science 172:*1229 (1971).

Table 36. Identifiable Species of Atmospheric CO in Rural Illinois
(*Source: Argonne National Laboratory*)

Variety	^{18}O Enrichment[a] (%)	^{13}C Depletion[b] (%)	Principal occurrence	Source	Production rate in northern hemisphere
AGA[c]	2.46	2.74			
1	0.5	3.0	Principal species everywhere. Increased abundance in summer	Methane	$>3 \times 10^9$ ton/year
2	0.5	2.4	In varying amounts with 1; increased concentration in winter and spring; also in marine air of low northern latitudes	Probably methane	
3	1.6–1.8	2.8	Lesser abundant heavy oxygen species during summer	Unknown	$\sim 5 \times 10^7$ ton/month during summer
4	2.6–3.3	2.2–2.6	Major species during autumn	Degradation of chlorophyll	2–5×10^8 tons during autumn
5	2.0–2.5	2.7	Major species during winter and early spring	Primarily anthropogenic	3–6×10^7 ton/month during winter

[a] With respect to the accepted oxygen isotopic standard, standard mean ocean water.
[b] With respect to the accepted carbon isotopic standard, Peedee belemnite.
[c] AGA, average global automobile.

6 – 1c   Methane and Natural Gas

Methane is present in the atmosphere as 2.2 ppm by volume, a concentration approximately one-hundred-and-fifty times less than that of carbon dioxide and about twice that of krypton, a "rare" gas. Methane is a hydrocarbon associated with petroleum wells, with natural gas wells, and with only a small amount of "petroleum liquids," carbon dioxide wells, salt wells, coal seams (fire damp) and coal mines, and decaying vegetable and animal matter. Methane is the major fuel component of natural gas and is transported under pressure many thousands of miles in the most intricate system in the world, the natural gas pipeline system. It is also known as "marsh gas" and burns in nature to give *ignis fati,* or "swamp fires." It is a colorless, odorless gas with a density of 0.7168° g per liter and a solubility of only 9 cc per 100 cc of water. The carbon monoxide produced by oxidation of methane to carbon monoxide and water amounts to 50 times that produced from internal combustion engines.

Methane results from the anaerobic decomposition of organic matter in swamps, marshes, ponds, lakes, slow streams, and oceans, and may be considered as a step in the carbon cycle. The chemical industries employ natural gas in great quantities as a raw material for the manufacture of ammonia, urea, and a host of nitrogen compounds. Methane is generated in garbage and rubbish landfills, and these occasionally blow up. The *Merck Index* gives the properties of methane:

> Marsh gas; methyl hydride. CH_4; mol. wt. 16.04. C 74.87%, H 25.13%. Widely distributed in nature. American natural gas is about 85% methane. The earth's atmosphere contains 0.00022% by vol. Major constituent of the atmosphere of the outer planets. (Jupiter, Saturn, Uranus, Neptune). . . . Colorless, odorless, non-poisonous, flammable gas. Burns with pale, faintly luminous flame. $d^{\prime}_4$ 0.554 (air = 1) or 0.7168 g/liter. m. −182.6°. b. −161.4°. Crit. temp. −82.25°; crit. press. 45.8 atm. Heat of combustion: 978 B.t.u./cu. ft. at 25° (a kilogram of CH_4 yields 13,300 kg.-cal.). Forms explosive mixtures with air, the loudest explosions occur when one volume of methane is mixed with 10 vols. of air (or 2 vols. of oxygen). Air contg. less than 5.53% methane no longer explodes. Air contg. more than 14% methane burns without noise. Autoignition temp. 650°. Solubility in water at 17°: 3.5 ml./100 ml. H_2O. Soluble in alcohol, ether, other organic solvents.

> Use: Constituent of illuminating and cooking gas, in the manufacture of hydrogen, hydrogen cyanide, ammonia, acetylene, formaldehyde, in organic syntheses.

> Human Toxicity: A simple asphyxiant. Narcotic in high concentrations in absence of oxygen.

Some 450 million years ago the Permian and Val Verdes basins in Texas were part of an inland sea, with the now-eroded mountains as islands and peninsulas. Here, a natural gas transmission company has drilled holes 20,000 ft deep through 450 million years of sedimentation and evaporation. The first deep well bottomed out at 23,140 ft and resulted in the discovery of the

Ellenburger zone — where the sea deposited its sediments, forming a layer of rock. The drilling operation took 6 months.

The Ellenburger zone is the deepest formation that has been drilled into in west Texas. This strata of rock is not porous, but internal earth movements have disrupted the Ellenburger, causing fractures. Hydrocarbons accumulate in these fractures, and provide reservoirs when discovered. Temperature and pressure at such depths are so great that hydrocarbons exist only as natural gas. The gas that is found is low in liquid condensate, with a measurable amount of carbon dioxide and hydrogen sulfide. Completion of the first well recorded natural gas at a calculated absolute open flow rate of 36 million ft^3 per day.

Trunk pipelines extending for 252,000 miles already exist in the United States for transmission of natural gas from areas where the gas is produced to areas where it is consumed. The system, which is constructed almost entirely of welded steel pipe, carries approximately 61.4 billion ft^3 (or 1.5 million tons) of natural gas per day.

6 – 1d   Cellulose and Coal

Coal results from the transformation of plant matter, whereby the oxygen and hydrogen in the woody fiber are eliminated in proportionally larger amounts than carbon.

Coal is believed to have originated in large marshes and swamps where the plant growth, including trees, was luxuriant. Huge accumulations of cellulose and lignin provided food for saprophtic fungi and bacterial fermentation. Oxygen and hydrogen were eliminated as water, carbon dioxide, and methane by the bacterial action which converted the material successively to peat and then to lignite. Bacterial action is responsible for the nitrogen in coal precursors. The conversion process was eventually ended by the formation, as a terminal product of bacterial action, of ulmic acid, which is 65.31% carbon and 3.85% hydrogen and is a sufficiently strong antiseptic to stop the action.

As a result of earth movement the swamp sank, and layers of inorganic sediments and sedimentary rocks covered the ulmic acid beds. Metamorphism, the reaction of heat and pressure, is thought to have been responsible for coalification. Wood, having become peat, was thereby transformed to lignite, which in turn became bituminous coal, and with a further increase in carbon and a reduction of hydrogen and oxygen finally became anthracite, as shown in Table 37. The rank of coal is a measure of the degree of metamorphism it has undergone, its age, and the effect of earth movements, the depth of rock layers above, and pressure from folding and faulting.

Coal then is the ultimate residue of another of nature's graveyards. Swamps and other areas where luxurious growth collected after death, where

Table 37. Composition of Fuels (in Percent)[a]

Fuel	Carbon	Hydrogen	Nitrogen	Oxygen
Wood	49.65	6.23	0.92	43.20
Peat	55.44	6.28	1.72	36.56
Lignite	72.95	5.24	1.31	20.50
Bituminous coal	82.24	5.55	1.52	8.69
Anthracite coal	93.80	2.81	0.97	2.72

[a] Source: CLARK, *The Data of Geochemistry.*

bacteria were nourished and perished in their own wastes, where the uneasy earth's crust formed covering layers of rock and earth all combined to complete the burial. We use the fossil remains as a fuel, and the resulting carbon dioxide is reused in photosynthesis for plant growth, beginning the cycle anew.

6 – 1e   Petroleum

Oil seeps have been observed by humans since Biblical times. Scholars suggest that Noah used natural tar to caulk the seams of his Ark and that Pharoah's daughter found the infant Moses in a papyrus boat protected by bitumen, or natural asphalt. "Greek fire," a weapon of the Bronze Age, was lighted petroleum skimmed from Middle Eastern waters. For numberless generations, the seepages of oil springs, found all over the world, have been used to caulk boats, preserve wood, heal wounds, cure disease, start fires, embalm bodies, waterproof articles, and cement stones.

On the west coast of Trinidad, an island in the Caribbean off the north coast of South America, there is a lake completely covered with asphalt. This has been the source for street-paving materials for New York and other cities, and is known as "Trinidad asphalt." In Trinidad, a submarine spring 200 yards offshore bubbles up, yielding brown oil and natural gas.

The La Brea tar pits, now part of the park system of Los Angeles, California, are so old that they contain the remains of extinct animals from the age of woolly mammoths and dinosaurs.

In 1793, the English explorer, Vancouver, was on a sailing ship near a promontory that was later named Coal Oil Point, off the coast of what is now Santa Barbara. Vancouver wrote in his log: "As far as the eye can see, the sea is covered with a sticky, smelly substance." Oil bubbled to the surface of the channel and spread in a slick widened by wind and current to the shore; it coated rocks and the sandy beaches, as well as unwary birds, observed by a sea captain 1 mile offshore. This was many years before we learned how to drill oil wells. The oil that Vancouver saw came from one of nature's oil wells, or seeps, which have been leaking oil and gas for millions of years. There are hundreds, perhaps thousands of seeps, found the world over, some on land, some on the sea floor.

Explorers have been noting the presence of oil seeps for centuries, and their observations have led to the discovery of important oil reserves. John Muir, a nineteenth-century naturalist and writer, described many oil and gas seeps in the Gulf of Mexico in his classic work, *Geology of the Tampico Region*. Later, geologists probing for oil in the Gulf consulted Muir's charts as a starting point in their exploration. John Muir is venerated as the "Father of Yosemite," one of the earliest national parks.

In the 1890s, de Morgan, a French geologist and archeologist wrote of seeps in the oil-rich Middle East. Maps of Iran are dotted with places named *naft,* or *naftun,* meaning "oil." De Morgan's work resulted in the first oil well drilled in the Middle East, at a place called Maidan-Naftan.

Oil seeps were noted in Alaska as early as 1896. A noted outdoorsman, hiking in the foothills near Katalla, reported smelling a "petroliferous odor, something like kerosine." He saw oil and tar floating on a river. At the source he found that "the entire area was covered with oil and a black asphalt residue."

Oil seeps have led to the discovery of most of the world's major oil fields. Oil seeping into Oil Creek led Samuel Drake to drill the first oil well in the United States at Titusville, Pennsylvania.

Sedimentary rocks are likely sources of seeps, involving formations that have been uplifted or disturbed by recent upheavals. Some seeps occur where deposits of oil have been broken open by movements, others where erosion has exposed oil-bearing rock beds.

The National Sea Grant Program of the National Oceanic and Atmospheric Administration is involved in a major seep study in the Gulf of Mexico. "Everybody complains about tar and oil in the water and on the beaches of the Gulf," says Dr. William Sweet, a geologist at Texas A&M University's Department of Oceanography, which is conducting the study. "Almost automatically the blame is placed on the offshore drilling rigs and tankers in the Gulf." One conclusion from this is that natural tar seeps are a prime source of pollution in some parts of the Gulf. Lumps of tar found floating in the northwest portion of the Gulf were found not to be refined products, but of natural origin.

Santa Barbara's oil seeps, which also predate drilling, were once valued as the source of a useful commodity. Seep tar was collected from beaches in the 1890s and was melted and shipped as far as San Francisco and New Orleans for use in street paving. Earlier, California's Chumash Indians used it to caulk canoes, to glue arrowheads to shafts, and to mend broken pots.

Following the 1969 Santa Barbara oil-rig blowout, Alan A. Allen, a professional oceanographer, used aerial, surface, and underwater investigative techniques to calculate rate of flow. The Coal Oil Point seeps, he found, leak 50–70 barrels of oil a day (2000–3000 gal) into the Santa Barbara Channel, and have probably been doing so for centuries. Using the lower, more conservative figure, Dr. Allen estimates that nature is putting around 18,000 barrels of oil into the channel per year. According to Dr. Weaver, professor of geology from the University of California at Santa Barbara, the Santa Barbara seeps are considered to be tens of thousands of years old. He has studied Santa Barbara for 15 years. He says that "the oil seeping from Coal Oil Point comes from a brittle and much-fractured rock formation, the Monterey shale, lying near the surface of the ocean floor at Coal Oil Point and readily leaks oil into the sea there and from the cliffs at the water's edge where it is exposed."

The Coal Oil Point seeps have their counterparts elsewhere along the California coast. One major seep off Redondo Beach regularly leaks oil onto the beaches of one of the state's most populous areas. Californians to whom the seeps are a nuisance have asked if they can be stopped, or if the oil could be cleaned up before it gets onto the beaches. Wells could be drilled to extract oil from the reservoirs that supply the seeps. Dr. Weaver has stated that a seep at More Mesa dried up when gas was withdrawn from the nearby La Goleta gasfield.

Engineers, using the Coal Oil Point seeps as a handy place to test clean-up devices, found that oil could be collected with booms and skimmers at a cost of $4000 a barrel. Thus, to collect 50–70 barrels of oil from the water's surface — the average daily production of the Coal Oil seeps — would cost $200,000–$300,000.

If nature pollutes the oceans, it also cleans them up. With oil seeping to the surface for centuries, and tar washing up on beaches daily, it would be reasonable to suppose that the beaches would soon be completely covered. But in time the oil and tar disappear of their own accord with no apparent harm to the environment.

Oil is a natural organic substance formed out of the remains of plants and animals, and is itself food to some kinds of organisms. Many microorganisms eat oil, and in digesting it eventually convert it into carbon monoxide, water, and other simple substances. Oil is rated as perhaps the least harmful, biologically, of aquatic pollutants, and oil seeps have been around for thousands, and probably millions, of years. Geologists believe: "We have never tried to turn off a volcano. They are there, and we have learned to live with them. Why shouldn't we do the same with nature's oil wells?"

In a practical sense, petroleum deposits are nature's burial grounds, the result of fundamental earth processes that have been operating since abundant and varied life was developed in the sea. The generation of petroleum from earth and sea involves the living and dying of organisms, the deposition of sediments, the advance and retreat of the seas, and the foldings of the earth's crust.

Conditions favoring petroleum production are represented by the stagnant waters of the Black Sea or of certain Norwegian fjords. The abundant life in the Black Sea is confined to the upper layers. The deeper and bottom waters are devoid of oxygen and often contain hydrogen sulfide from the decomposition of sulfoproteins from animal tissue. In these waters there can be no bottom scavengers to devour the bodies of marine animals that drift down from above, which are therefore entombed in the fine sediments. In many Norwegian fjords the deep layers are without oxygen. The mouth of the fjord is cut off from the circulation of the open sea by a shallow sill. Storms sometimes drive in unusual quantities of oceanic water and through the turbulence of waves deeply stir the waters of these pools. The mixing of the water layers brings death to hordes of fishes and invertebrates living near the surface, which leads to the deposition of a rich layer of organic materials on the bottom.

The search for petroleum has led geologists to areas formerly covered much of the time by shallow seas. These lie around the margins of the continental shelves, between them and the oceanic deeps. There is a depressed segment of crust lying between the continental masses of Europe and the Near East, consisting of the Persian Gulf, the Red, Black, and Caspian Seas, and the Mediterranean Sea. The Gulf of Mexico and the Caribbean Sea lie in another basin of shallow seas between North and South America. A shallow, island-studded sea lies between the continents of Asia and Australia. There is also the nearly landlocked sea of the Arctic. In past ages all of these areas have been alternately raised and depressed, and all were at one time part of the land, and at another time part of the sea. During submersion they received thick deposits of sediments; in their waters marine fauna lived, died, and drifted down into the soft sediment carpet.

There are seeps, natural springs, and oil deposits in these areas. In the Near East are the oil fields of Saudi Arabia, Iran, and Iraq. The shallow depression between Asia and Australia yields the oil of Java, Sumatra, Borneo, and New Guinea. The American "Mediterranean" is the center of oil production in the Western Hemisphere. The oil resources of the United States come from the northern shores of the Gulf of Mexico, Colombia, and Venezuela. Mexico has oil fields along the western and southern margins of the Gulf. Oil seepages in northern Alaska, on the islands north of the Canadian mainland and along the Arctic Coast of Siberia, have been discovered recently.

Perhaps to the twentieth-century geologist, oceanographer, or petroleum engineer, the discovery of oil fields and the manufacturing of refined petroleum products is the practice of modern recycling, in terms of geologic time. There never was a pristine earth, stable, always beautiful, peaceful, and clean; there has always been change, destruction, and catastrophe. As temporary inhabitants, we have to maintain and perhaps improve our housekeeping, a job well advanced by actual doing, not by shouting about it. Man has to accommodate himself in terms of geologic time.

Beginning in 1894, when oil was first drilled from piers in shallow coastal waters, the sea's oil deposits have been increasingly searched for and tapped. This undersea petroleum was formed from the innumerable tons of sea plants and animals that have died and settled to the ocean floor since aquatic life began millions of years ago. Stored as a liquid or gas in the pores of sedimentary rock, the crude ranges from a thick black liquid to one that looks and pours like ginger ale.

Until 1947, marsh and swamp wells were the only wells in coastal waters. In that year, however, the first holes were drilled beyond the sight of land, 10½ miles from land in the shallow waters of the Gulf of Mexico. While those holes were sunk in barely 18 ft of water, exploratory wells have since been drilled in water more than 50 times that deep.

In the search for salt-water oil, aircraft are used to perform preliminary explorations. As they fly above the water, a magnetometer on the plane examines the physical properties of the rocks below the sea floor. When this probe has identified a promising area, it is checked more closely with a shipboard reflection seismograph. By measuring the shock waves reflected back to the surface from formations below the sea floor, the seismograph defines the structurally high areas where oil and gas fields are found.

Geologists on exploration ships seek supplementary data in a promising area by drilling holes in the upper layers of sedimentary rocks. An analysis of the data often indicates whether suitable conditions exist for hydrocarbons to be present in commercial quantities thousands of feet below.

After all these steps have been taken, a wildcat well is still the only way to prove the presence of a commercial deposit. Since 49 out of 50 wildcat wells prove to be dry holes — either literally or commercially — the exploratory rigs that drill these holes are usually mobile.

How much oil could there be in the billion billion tons or more of water that make up the world's oceans? Nobody knows. Tankers collecting samples while going about their regular business have recently sent back some 800 specimens. Each ocean sample is subjected to a series of procedures and tests that separate other organic chemicals from hydrocarbons and measure the amount and composition of the hydrocarbons present. The entire process has been refined to the point where one part of oil per billion parts of water can be detected.

The first 300 samples, which were taken on the trade routes linking the East Coast of the United States with the Gulf of Mexico and the Caribbean, showed an average oil concentration of about 6 ppb, an amount that could not be detected by older methods of analysis. It is hoped that these will provide a better idea of the amount of oil in the sea, how long it has been there, and how much of it arises from human activities rather than from natural sources.

The mistral wind blows from the Alps in the summer and sweeps through the Rhone valley in France to the warm beaches of the Mediterranean. It blows with a steady rush of gale force, bending cypress trees, carrying dusts and sand on the littoral from the Côte d'Azur to the Languedoc. After 3, 6, or 9 days, it stops. If it were harnessed by windmills it could create power. However, the power flow is in the reverse direction, in the form of crude oil moving 6–7 mph in one of the earth's major pipelines, from the Mediterranean to the industrial cities of Europe.

In a speech before the American Association for the Advancement of Science, Rene Dubos, of the Rockefeller Institute said:

> Ever since the beginning of the agricultural revolution during the Neolithic period, settlers and farmers have been engaged all over the world in a transformation of the wilderness. Their prodigious labors have progressively generated an astonishing diversity of man-made environments, which have constituted the settings for most of human life. A typical landscape consists of forested mountains and hills serving as a backdrop for pastures and arable lands, villages with their greens, their dwellings, their houses of worship, and their public buildings. People now refer to such a humanized landscape as "nature," even though most of its vegetation has been introduced by man and its environmental quality can be maintained only by individualized ecological management.

This describes the countryside of the buried pipeline in France. The pipeline extends for 489 miles, from the port of Lavera, near Marseilles, to Karlsruhe, West Germany, traversing French territory nearly all the way. Crude brought by tankers from North Africa or the Persian Gulf is fed through the pipeline to 11 refineries which supply petroleum products to the cities and factories of the upper Rhine and Alsace-Lorraine.

The Continent once relied on the coal deposits of the Ruhr and northern France for its basic fuel. Petroleum now supplies 60% of its energy. If the pipeline were not there and if the oil were shipped by railroad, 2000 tank cars going up the Rhone valley each day and coming back would be needed.

The pipeline runs through the valleys of the Rhone, Ain, and Doubs rivers, through the Belfort Gap between the Vosges and Jura mountains, and finally across the Alsatian plain to the Rhine. This has been a prime trade route since the time of the Phoenicians. The route does not present spectacular physical obstacles (its maximum elevation is 1500 ft), but there were plenty of problems for the builders. The pipeline crosses some three dozen sizable rivers. It tunnels under a succession of highways, railroads, water conduits, and other barriers, with one such obstacle every 1500 ft. In the South, the pipeline runs beneath fields of lavender and the stubby, sunbaked vines that produce the rosé wines of Provence. After Avignon, it traverses the full 100-mile length of the Côtes du Rhone region, where the red wines rank with the best anywhere. It then crosses the dairyland of the Jura region, with its goat cheeses and the wine of Arbois. Further north, the pipeline passes beneath trellises of hops and mile after mile of the tall Alsatian vines of the delicate Rieslings, Sylvaners, and Gewurztraminers.

Over most of the route, the scars of pipeline construction have faded. There is nothing to show that crude oil is rushing beneath the fields at upwards of seven barrels per second. Yet the very richness of the land led to difficulties in negotiating for the right-of-way. This is one example of humanization, human adaptation, without any destructive effect.

6 – 2   THE INORGANIC ELEMENT CYCLES

6 – 2a   Nitrogen

Nitrogen, the major component of the atmosphere, is generally thought of as nonreactive, but human ingenuity has learned how to fix nitrogen (e.g., as ammonia, urea, nitrates, and amines) for fertilizer in greater amounts than are fixed by the roots of legume plants and associated soil bacteria through natural biological nitrogen fixation.

At higher than normal pressures and temperatures, as well as in the presence of a catalyst, nitrogen reacts to form oxides of nitrogen and nitric acid, the basis of most of our industrial and military explosives. Environmentalists worry about the oxides of nitrogen from the exhausts of internal combustion engines.

Oxides of nitrogen may be formed in the atmosphere by lightning, corona-power effects, and the like. Their water solubility is high, so that they are carried down by rain to add to soil fertility.

Nitrogen in fixed form is essential to the growth and life of green plants and is essential for the formation of proteins necessary for animals and humans. When proteins in dead plants or animals decompose, ammonia or related amines are formed and recycled. Trees fix nitrogen indirectly, as shown in Figure 27.

6 – 2b   Sulfur and Sulfur Dioxide

Sulfur is an essential constituent of certain proteins, termed sulfoproteins, particularly in animals and fish. In recycling or decomposition, hydrogen sulfide is formed, a colorless gas that smells like rotten eggs and blackens many metals by forming dark-colored sulfides. Sulfur is associated with coal, petroleum oil, lignite, and peat. When these are employed as fuels they give rise to sulfur dioxide. From the standpoint of the atmosphere, sulfur dioxide is the most important sulfur compound. A large number of metallic ores of copper, lead, and zinc are sulfides, i.e., they are combinations of the metal and sulfur.

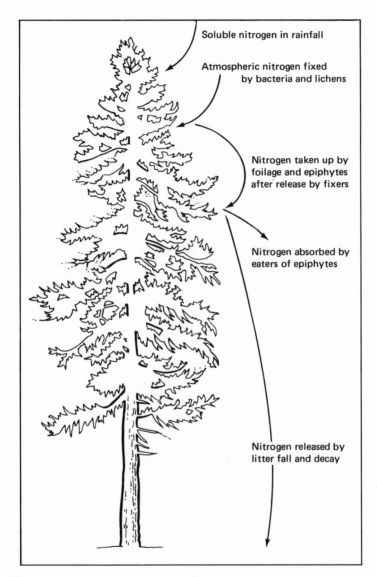

Soluble nitrogen in rainfall

Atmospheric nitrogen fixed
by bacteria and lichens

Nitrogen taken up by
foilage and epiphytes
after release by fixers

Nitrogen absorbed by
eaters of epiphytes

Nitrogen released by
litter fall and decay

FIGURE 27. Nitrogen pathways in the forest. Some nitrogen (less than 1 lb per acre annual-ly) is brought by rain. More enters through nitrogen-fixing bacteria on the fir needles and on blue green algae in some lichens or moss. Rain leaches nitrogen from dead tissue. The excretion of herbivorous animals and death both add more nitrogen to the soil.

From the viewpoint of the chemist, sulfur dioxide's properties are relatively innocuous as compared with those of carbon dioxide.

A comparison of the properties of carbon monoxide and sulfur dioxide shows a solubility in water (including seawater) of 3.3 volumes per 100 volumes for CO, and a greater solubility of 17.7 volumes per 100 volumes for SO_2, which is converted to sulfur and sulfates.

The maximum allowable concentration for CO is 100 ppm, while the lethal concentration is 1000 ppm. Oxygen is the antidote. There is no maximum allowable concentration for SO_2, and the lethal concentration to rats in four hours is 1000 ppm, when SO_2 is employed as a fumigant.

According to EPA data, only 8% of the CO in the atmosphere is from man's activities and 92% is from natural forces. For SO_2, the figures are 67% from volcanoes and 33% as a result of man's activity.

The properties of sulfur dioxide, as given in the *Merck Index,* are:

Sulfur Dioxide: Sulfurous anhydride; sulfurous oxide. SO_2; mol. wt. 64.07. S 50.05%, O49.95%.

Colorless, nonflammable gas; strong suffocating odor. Condenses at $-10°$ and ordinary pressure to a colorless liquid. d. gas 2.3; d. liquid 1.5. Solidif. $-72°$. b. $-10°$. Mixed with oxygen and passed over red-hot Pt is converted into SO_3. With water forms sulfurous acid (H_2SO_3). Bleaches vegetable colors. Solubility: in water 17.7% at 0°, 11.9% at 15°, 8.5% at 25°, 6.4% at 35°; in alcohol 25%; in methanol 32%; also soluble in chloroform or ether. It is supplied compressed in cylinders. Lethal concn. to rats in air: 1000 p.p.m.

Use: Preserving fruits, vegetables, etc.; disinfectant in breweries and food factories; bleaching textile fibers, straw, wicker ware, gelatin, glue, beet sugars.

Human Toxicity: May be intensely irritating to eyes and respiratory tract.

Sudbury, Ontario, Canada, is a monument to heap-leaching on the ground of sulfide ores, which converts them to oxides and sulfur dioxide. Within a 25-mile radius there is practically no vegetation, merely thousands of dead tree stumps left over from widespread timber-cutting of the late nineteenth century. Whenever it rains, mucky silt washes into the city streets from the denuded hills. Dust-puffing smokestacks create a reddish-brown pall over the area.

Before modern smelting techniques came into use in the 1930s, outdoor roasting was used to remove sulfur from nickel ore. This method consisted of piling the ore onto huge outdoor wood fires, producing great amounts of sulfur-bearing smoke that killed off most vegetation and more or less petrified the thousands of tree stumps that already existed. Since the 1930s, furnaces, instead of outdoor fires, have churned out the gases that cause fumigations and other problems. The International Nickel Company at Sudbury was ordered to cut its emissions 85% by 1978, and, as an interim measure until the emission level declined that far, to build what turned out to be the world's largest smokestack, a 1250-ft-high, $26 million structure designed to disperse the gas over a larger area.

A similar deforestation took place at Copper Hill, Tennessee, but the Tennessee Copper Company is now the largest sulfuric acid producer in the world, because they catch the sulfur dioxide and convert it to acid. The Copper Hill environs are being reforested.

O'Neal and Kroetz reported[5] that tall stacks contribute very little to ground-level sulfur dioxide. The Long Island Lighting Company (LILCO) has operated an extensive monitoring system for over 3 years. Nearly 900,000 separate pieces of information have been recorded for a land area of some 600 square miles, 15 miles north-to-south by 40 miles east-to-west. The data consist of approximately 500,000 sulfur dioxide measurements at various locations, 160,000 pieces of weather data, and 220,000 measurements of sulfur dioxide as emitted from each of the power-plant stacks on the system, all measured on an hourly basis and stored on computer cards and tapes. The computer has been programmed to collate stack emissions, weather information, and ground-level sulfur dioxide measurements. The analyses show that emissions from tall stacks do, indeed, contribute very little ground-level sulfur dioxide. For example, LILCO shows that its Northport plant contributes less than 2 ppb on an hourly annual average. Conversion of the plant to 100% natural gas, or shutting it down, would therefore reduce ground-level sulfur dioxide by no more than 2 ppb per hour.

It is often heard that of the poisonous gases released into the atmosphere the worst is sulfur dioxide, whose build-up is supposedly increasing constantly. But data gathered by EPA's National Air Pollution Control Administration in 30 major cities from 1964 to 1969 show that, in 1969, 17 of those 30 cities had a lower level of SO_2 than 5 years earlier, and three cities had the same level; and of the 10 cities exhibiting a higher level, only five (New York, Boston, Providence, Philadelphia, and Dayton) had a SO_2 concentration greater than 0.04 ppm.

The National Center for Atmospheric Research found that the half-life, or decay, of SO_2 in the atmosphere is somewhere between 4 hours and 14 days. Studies at the Stanford Research Institute showed that "SO_2 lasts about four days in the atmosphere, being gradually converted to sulfate produced by combination with traces of ammonia always present in the air." Ammonia is one of the many waste products of plant and animal metabolism. The worldwide ecological cycle removes naturally produced sulfur dioxide.

Stoiber and Jepson[6] made extensive measurements, by remote sensing equipment, of Central American volcanoes. Quiescent volcanoes were estimated to produce 100 tons of SO_2 per day, an amount larger than that produced

[5] A. J. O'NEAL JR., and C. A. KROETZ, *Transactions of the Society of Mining Engineers, AIME,* 252:399 (1972).
[6] R. E. STOIBER and A. JEPSEN,

by the average coal-burning power-generating station. The total amount of SO_2 from Central American volcanoes is 1300 tons per day.

An emission of 100 tons per day of SO_2 would be equal to the removal of sulfur from a 1-million-ton ore body with 50% sulfur in 25 years. This is the equivalent of processing 40,000 tons of iron pyrites per year.

The minimum estimate of the annual amount of SO_2 emitted from the world's volcanoes is about 10^7 metric tons.

Kellogg et al.[7] indirectly estimated the annual contribution of SO_2 pollution as 10^8 tons. None of the volcano SO_2 is collected, but much of the SO_2 from ore processing is collected and used, so that the inference may be made that volcanoes are now or will be the major source of SO_2 in the atmosphere. Industry is not the dominant agent of SO_2 emissions. Volcanoes and the oceans produce 67% of the sulfur oxides present in the atmosphere, while all human activities produce only about 33%.

Decreases in the concentrations of particulates and lower atmospheric SO_2 concentrations have their origins in technology. In the early decades of this century wood- and coal-burning fuels were replaced by fuels such as oil and natural gas. Later, better and more efficient particulate removal, by scrubbers, electrostatic precipitators, and filters, became the rule, and SO_2 recovery systems were developed over the years.

The view of the engineer has been well expressed by Eugene Guccione, senior editor of *Engineering and Mining Journal,* New York, in the June, 1971, issue:

> The true "environmentalist" is the scientifically oriented individual who applies knowledge and reason to the problems of environment. Such an individual is not to be confused with the self-styled "environmentalist" whose motives vary from blind idealism to outright self-aggrandizement, and who compensates for his lack of knowledge and rationality by his vociferousness.
>
> Environmentalists cling to the notion that SO_2 concentration in the atmosphere increases annually. If that were the case, none of us would be here today. Our ancestors would have died — suffocated by the catastrophic buildup of SO_2 which began when Prometheus brought the fire of the gods to men.

6 – 2c Phosphorus

Phosphate rocks, which derive from the skeletal remnants of animals, are mined by man, made soluble by sulfuric acid, and converted into fertilizers for plants. Fertilizers are graded according to their potassium, nitrogen, and phos-

---

[7] W. W. Kellogg, R. D. Cadle, E. R. Allen, A. A. Lazrus, and E. A. Martell, *Science, 175:*587 (1972).

phorus contents. Plants form phosphoproteins, and when vertebrates eat the plants, the phosphorus compounds help to form skeletal members, which are primarily composed of calcium phosphate. When the animals die they leave behind the mineral constituents of phosphate rocks, apatite and fluoroapatite.

6 – 2d   Calcium

The calcium cycle, associated with carbon dioxide, is found in carbonate rocks (along with magensium). Calcium carbonate is only slightly soluble in water, but is markedly soluble in water saturated with carbon dioxide. Thus bicarbonates are formed, and the water is termed "hard." Mollusks convert calcium compounds with carbon dioxide into protective-shell material (e.g., oysters, clams, and scallops) and all animals need calcium for skeletal materials.

In the carbon dioxide cycle, calcium forms marble, sedimentary limestones, and other rocks. Calcium is almost as abundant as iron, being the fifth most abundant metal.

6 – 2e   Metals

Trace concentrations of certain elements, such as iron and zinc, are essential to human metabolism; other elements, such as lead and cadmium, are toxic at similar concentrations; and still others, such as selenium, can be either beneficial or toxic within a fairly narrow range of concentrations.

Lead and methyl mercury have been shown to damage the central nervous system.

Beryllium is carcinogenic, and is a short-term poison in high concentrations and a long-term systemic poison in low concentrations. Nickel carbonyls have been implicated as a source of lung cancer, and cadmium, arsenic, selenium, and yttrium have been shown to be carcinogenic in laboratory animals.

Many trace elements are, of course, beneficial. At the latest count, 14 different trace elements have been identified as essential to human health. Cobalt, zinc, and manganese serve as cofactors for various metabolic enzymes, and iron is an integral component of hemoglobin. Trace concentrations of fluoride in water have assisted in the prevention of dental caries.

Cadmium has a pronounced effect on bones, according to Jun Kobayashi of Okayama University, Kurashiki, Japan. Kobayashi had previously established a relationship between the mysterious syndrome called itai-itai ("ouch-ouch") disease — which affected residents of Japan's Jinzu River basin in the late 1950s — and the discharge of cadmium into the river by a large factory. Ingestion of cadmium by the residents, he suggested, led to the severe and painful decalcification of skeletal bones and multiple pathological fractures characteristic of the syndrome.

Many other instances of poisoning by trace metals from natural or man-made sources will undoubtedly be discovered as scientists extend their investigations of mysterious illnesses.

Natural sources of trace elements may also produce unexpected effects on health. The Pima Indians of the Gila River Reservation in Arizona provide an example of these effects. Among Pimas who have spent most of their lives on the reservation, the incidence of obesity, cholesterol gallstones, cirrhosis of the liver, and diabetes mellitus (45% of adult Pimas are diabetic) are substantially higher than in the population at large, and the incidences of duodenal ulcers, hypertension, arteriosclerosis, and cancer of the lung and breast are substantially lower.

Domestic water on the reservation is unusually enriched in several elements, including sodium, sulfate, strontium, boron, lithium, and molybdenum, and is deficient in the required elements copper, zinc, and manganese. Certain food plants also tend to concentrate particular elements. For example, mesquite beans accumulate strontium, cabbages accumulate sulfate, beans concentrate molybdenum, and wolf-berries (used for jelly) contain an extraordinary amount of lithium (1120 ppm).

Despite the growing body of knowledge about trace elements and their association with illness, there has been little application of research findings to clinical medicine.

6 – 2f Chlorine

Chlorine constitutes 314 ppm of the earth's crust. In the United States the chemical industry produces 30,000 tons per day from salt, half of which is consumed in making other chemicals. Man has mined and produced salt from the sea in small quantities essential to his use; he cannot ingest seawater in that it contains salt. Chlorine and its compounds protect man as antiseptic agents against pathogenic organisms. In the United States, chlorination of drinking water from 1–2 ppm to as high as 10 ppm is almost universal. The rest of the world does not pay that much attention to its potable water. As a result, water-born human diseases are rare in the United States.

Animals will travel miles to salt licks since they, like man, need to maintain a physiologic salt balance.

6 – 2g   Silicon

Silica, the oxide of the element silicon, is the most abundant element on earth and is generally thought of only in connection with rocks (e.g., silicates), beaches (e.g., quartz sand), and inorganic cycles. However, diatoms, radiola, and similar marine organisms develop skeletal structures of silica, and immense deposits of diatomaceous earth, the burial grounds of these organisms, are found in areas that were once covered by the sea.

7

HUMANIZING THE EARTH

7-1 AGRICULTURAL ENGINEERING

The use of lime on acid soils is a necessary part of farming programs on most humid-region soils. Liming became a general practice in this country around 1900. Twenty-three million tons of lime were applied in 1957. An estimated 80 million tons of liming materials are needed annually to maintain soil fertility and permit maximum crop yields.

Liming is particularly important because when soils become too acid, stands of legume-forage fail, growth of other crops is inhibited, and fertilizer nutrients revert to insoluble forms that the plants cannot use. The increased use of nitrogen fertilizers is increasing soil acidity. It takes about 550 lb of limestone to neutralize the acids that result from a 550-lb application of ammonium sulfate fertilizer.

Natural organic materials formerly supplied 90% of the commercial nitrogen. They now supply less than 3%, a consequence of the discovery of methods of fixing nitrogen from the air through the union of hydrogen and nitrogen to form ammonia. Large-scale manufacture of ammonia opened the way for low-cost nitrogen materials, such as ammonium nitrate, urea, aqueous nitrogen solutions, and anhydrous ammonia. The only other large sources of commercial nitrogen are ammonium sulfate, produced as a byproduct of the coke and steel industries, and South American deposits of sodium nitrate.

A great advance in the preparation of phosphate fertilizers occurred more than 100 years ago with the manufacture of superphosphate, which still is the most widely used commercial carrier of phosphorus. After scientists proved that mineral phosphates were suitable for making superphosphates, explorations disclosed widespread and almost inexhaustible phosphate deposits, some of the largest of which are in the United States.

Another important capital resource that can partially substitute, so to speak, for land to meet production needs, is the rapidly expanding group of chemical weedkillers and insecticides. In 1957, farmers spent $231 million for agricultural chemicals other than fertilizer. Weeds cause losses on American farms that are estimated to run to $5 billion annually. Weeds compete with crops for water, nutrients, and light. They increase the cost of labor and equipment and reduce the quality of farm products. A ragweed plant is said to need three times more water than a corn plant. One plant of common mustard takes twice as much nitrogen and phosphorus, and four times as much potassium and water, as an oat plant. Weeds also add to costs of tillage and seed cleaning, give milk off-flavors, reduce the quality of grains, and make harvesting difficult.

Several chemicals have been produced to kill nematodes. Dichloropropane and ethylene dibromide have been used for a number of years, and, more recently, 1,2-dibromo-3-chloropropane has been introduced. These chemicals are effective and are used widely on tobacco, vegetables, and pineapples. They are too costly for many crops.

One generation ago, grasshoppers could be partially controlled by applying 20 lb of poison bait per acre of crop. Two men could cover perhaps 150 acres per day. Now, 1000 acres can be sprayed with chlorinated hydrocarbons from an airplane in a few minutes, and better protection is provided. Ground and aerial equipment for applying chemicals is being improved to give more thorough treatment with smaller amounts of insecticide.

But the process is not as simple as it might seem. As growers strive for greater and more efficient production of cotton, insect problems often become more acute. When cotton is grown under irrigation using large amounts of fertilizer, the larger plants and the longer growing and fruiting periods that result favor a build-up of boll weevils and other pests. The wise use of insecticides throughout the growing period is essential to obtain the extra yields that such growing practices provide.

The control of livestock pests reduces loss by death, improves animal health, and increases the efficiency with which feed is converted into animal products. As a result, fewer acres are needed to produce the feed needed for a given number of pounds of meat, milk, or wool.

Many technical improvements are being made in storage and processing facilities for use on and off the farm. About 119 million bushels of stored cereal were ruined by insects in the United States in 1952. As much as 10% of the stored wheat in the Great Plains can be destroyed in a season. Insect damage to corn stored in the South under farm conditions may be 9% per month in bad seasons.

Experimental work has been done on the use of antibiotics on stored food products to reduce spoilage, enhance quality, and extend shelf life. Promising results have been obtained with some of the tetracycline compounds and a few other antibiotics, in experiments conducted with fish, milk, and eggs.

The patrons of organic gardening and organic foods are often highly vocal and single-minded, as well as being affluent consumers of higher-priced foods than others can afford.

Fertilizers and lime may have contributed more than any other technological advance to sustained production and efficient use of land. Without fertilizers and lime, intensive farming would no longer be profitable in many areas that are now intensively farmed.[1]

The need for plant foods has been recognized for centuries, but chemical fertilizers have come into use only in the past 100 years. The American fertilizer industry produced a few hundred tons in the 1850s, 22 million tons in 1957, and has produced larger amounts each year since.

Crop losses due to pests and disease have plagued man from the beginning of history. The word "pest" includes birds, insects, mites, mollusks, nematodes, rodents, plant pathogens, and weeds. Plagues have affected man's health for many centuries.

Accounts of the losses caused by man's major food competitors are found in the writings of classic Greek and Roman authors, among whom Theophrastus, Virgil, Pliny the Elder, and Columella are probably the most representative. So great was the fear of crop losses that pests and diseases and their causal agents became an integral part of superstitions and dogmas in most of the early civilizations. In the pagan liturgy of the Romans, special rites called "Robigalia" were introduced in April of each year to propitiate the goddess Robigo. This divinity, which was identified with cereal-rust disease, was considered to be so important during the spring that she was recognized as the "maxima segetum pestis" (the worst crop pest) if not properly pacified.

A swarm of the desert locust, *Schistocerca gregaria*, can cause tremendous damage. Each locust nymph is capable of eating its own weight in food every day, and swarms of 100–200 million insects per square mile, covering sometimes up to 400 square miles, are capable of destroying 80,000 tons of crop per day.

Crop losses caused by pests and diseases have been so great and so frequent that they have actually led to famine conditions, and more than once have contributed to great changes in many countries.

Pesticide costs have increased considerably, placing much of the financial burden on consumers of farm products and on farmers. It has been estimated that while in 1955 the cost of pesticides in the United States was about 1% of the total crop value, in 1968 this had risen to 4.6%. Many different causes have contributed to this sharp increase. For example, the immediate and often spectacular results obtained with pesticides, combined with consumer demand

---

[1]*Land: The 1958 Yearbook of the Department of Agriculture*, U.S. Gov't Printing Office, Washington, D.C., 1958.

for pest-free, blemish-free products, have required increased pesticide usage and have broadened the spectrum of treated crops. At the same time, the development of pest-resistance and accompanying pest-resurgence problems has encouraged the use of either higher pesticide concentrations or more expensive substitute chemicals. Costly pesticides were also introduced in many pest-control programs to replace cheaper materials which, for varied reasons, had been banned from common use. For the control of weeds, the farmer, faced with the increasing scarcity of labor or its increasing costs, had no other alternative but to use larger amounts of herbicides to raise his crops.[2]

S. C. Salmon, O. R. Mathews, and R. W. Leukel, of the Department of Agriculture, reported that most wheat varieties grown before 1900 have disappeared. Between 1900 and 1950, 284 new varieties were developed and grown on farms. The newer varieties have resulted in substantial increases in yield. At the same time quality has been improved, losses from shattering and lodging have been reduced, and production has been stabilized by reducing losses from winter killing, diseases, and insect pests.

Is this an example of elimination of species? They also say:

> Diseases threatened the nation's oats crop three times in 1938–58. New varieties have been developed that resist the diseases and are of higher quality and better yield.
>
> A promising start has been made on breeding resistance to diseases and pests into plants. An early achievement was the development of Pawnee wheat, which is resistant to loose smut and is widely grown in the winter-wheat areas of the Great Plains. The Indiana Agricultural Experiment Station has produced Dual, a soft-red winter wheat that has resistance to hessian flies and to leaf rust. The same station has also produced La Porte wheat, which is adapted to northern Indiana and is resistant to loose smut.

The average feed consumption per pound of broiler produced has dropped from 4.25 lb in 1940 to less than 3 lb in 1970. In experiments, broilers have been produced on less than 2 lb of feed per pound of broiler, live weight. This gain has resulted from improvements in breeding, feeding, sanitation, and management.

Experiments show that the use of stilbestrol, a hormone, in rations for fattening beef cattle will reduce the feed required per pound of gain by 10–15%. The addition of antibiotics to the rations of young pigs can increase the growth rate 10–20%, and can increase feed conversion efficiency as much as 5%. Antibiotics are also of value in the rations of young dairy calves.

Urea, a synthetic nitrogenous compound, can be used to supplement feeds produced from the land. Added to the rations of ruminants, it becomes a source of food for the microorganisms in the rumen. The bodies of the microor-

[2]*The Locust Handbook*, The Anti-Locust Research Center, London, 1966. (38056) 37.

ganisms in turn provide proteins for the animal. One pound of urea plus 6 lb of grain will replace approximately 7 lb of oil meal in the rations of cows, sheep, or goats. Urea can also be used to improve the feeding value of low-quality roughages. Urea is toxic in large quantities. Not more than one-third of the total nitrogen in the diet should be supplied by urea. The quantity fed must be carefully controlled and thoroughly mixed with the ration. Most of it is fed in ready-mixed formula feeds. In 1957, 87,000 tons of urea were fed, permitting the replacement of almost 610,000 tons of oil meal as a source of protein.

Mechanization has reduced labor requirements for many cultivated crops relative to production and feeding of hay and forage. And the production of abundant, low-cost inorganic nitrogen has reduced farmers' dependence on green manures.

While the rest of the world progresses to avoid Malthus' prediction that man's food production capacity would not be able to feed the increasing billions of people, the organic convert would like to regress.

Dr. Norman E. Borlaug, winner of the 1970 Nobel Peace Prize, made a speech in Rome before the biennial conference of the United Nations Food and Agriculture Organization. A plant pathologist, Dr. Borlaug won the Nobel Prize for his contribution to the development of high-yield wheat strains. Dr. Borlaug said:

> The continued success of the Green Revolution [a program which enables develop-
> ing countries to substantially increase their food production] will hinge upon whether
> agriculture will be permitted to use the inputs—agricultural chemicals including
> chemical fertilizers and pesticides, both absolutely necessary to cope with hunger. If
> agriculture is denied their use because of unwise legislation that is now being
> promoted by a powerful lobby group of hysterical environmentalists—who are
> provoking fear by predicting doom for the world through chemical poisoning—then
> the world will be doomed—not by chemical poisoning but from starvation.

Without thinking, conservationists and environmentalists, both in and out of government, and only partially informed people in the communications media, have embarked on a crusade designed to end the use of agricultural chemicals such as pesticides and fertilizers. They give no thought to the end result of such action: eventual world starvation and political chaos.

Corn has been credited with being the New World's secret weapon and the builder of its civilization, for it was not known in the Old World of Europe. In Mexico's Yucatan Peninsula, atop pyramids, courtyards, temples, and altars, more than a millenium before the discovery of America, there were carvings of ears of corn or some symbol of its cultivation. Similar monuments are found in Central America, Peru, Bolivia, the southwestern United States, and the Mississippi Valley. The Olmecs, the Mayans, the Incas, the Toltecs, the Aztecs, the Zapotecs, the Cliff Dwellers, and the Mound Builders, all depended on corn.

In the United States, corn is the top crop, about double that of wheat. Dr. Paul C. Mangelsdorf, a retired Harvard professor of botany, who has spent nearly all his adult lifetime studying the plant, has said "There is nothing like it either in nature or among cultivated plants." But the ear of corn is a paradox, for it cannot live without human help.

The corn of our primitive ancestors had cobs about the size of a stubby pencil that contained some 50 tiny kernels and were supported by a short, spindly stalk. Modern hybrids have ears as long as a man's forearm, bearing more than 1000 large kernels. Plant breeders have made corn higher in protein (field corn), higher in sugar (sweet corn), disease resistant, insect resistant, mold resistant, drought resistant, and wind resistant. Plants have been bred to bear ears at a uniform height for easier mechanical harvesting. Farmers plant them with a density per acre that allows each one to absorb the maximum amount of sunlight.

Improvements in modern corn have come from using ancient types as the breeding stock. The major types of corn—dent, flint, flour, pop, and sweet corn—were in existence 500–1000 years ago. When some of the ancient strains were inbred, their vigor decreased. But when separate inbred lines were crossbred, there was an explosive burst of hybrid vigor. Between 1929 and 1969, average corn yields in the United States changed from 26 bushels to 80 bushels per acre.

Henry A. Wallace, a corn breeder, Secretary of Agriculture, and Vice President under Franklin D. Roosevelt, introduced hybrid corn seed commercially in 1926. Before then, one farmer could feed about nine people. Today, a corn-belt farmer can grow enough corn to produce livestock products that will feed about 400 people (it takes 10 lb of corn to raise 1 lb of steak). Of course, mechanization, fertilizers, and pesticides aid in this production, but corn now has the capacity to repay all these inputs.

Man has searched for corn's beginning, for if geneticists could identify corn's predecessors, they would have important new gene pools with which to improve modern hybrids. One example of a new variety is "prolific" corn. Under good conditions it produces some 20% more than other varieties by growing two or three ears on a plant (some modern hybrids have only one large ear). Under stress conditions, such as drought, this corn still manages to produce one ear, whereas other varieties grow none at all. This corn was bred in part from an "archaic" South American popcorn.

Man's need for new genes came with the corn blight of 1970, which reduced the nation's corn production by some 700 million bushels. The devastation resulted from a gene pool that was too narrow. The corn was so similar that when a damaging disease came along, virtually every field in the country was susceptible. Here was a case of carrying all our corn, genetically speaking, in one basket.

The Federal Council for Science and Technology noted that, in the 1960s, agricultural science was ranked low by academic standards, but it nevertheless undergirds an efficient scientific agriculture. It has been stated that agricultural research is controlled by administrators rather than by peers in the laboratories.

The agricultural research system is finely attuned to the immediate needs of its clients. This fine tuning is evident in the response of the system to the attack of corn blight in 1970. Within a year, the blight was controlled and corn and meat were saved. The credit is shared by farmers, seedsmen, and scientists. In the 2 years before the epidemic there were only six publications on the subject; in the year of the blight there were 18; and in the 2 years since there have been 91 publications. In the short time since the epidemic, 62% of these reports have come from experiment stations using their own resources, and 22% more have come from the U.S. Department of Agriculture, either alone or in conjunction with the experiment stations. Clearly the response of agricultural researchers to a problem of their clients and of the nation was effective. Not all systems are so well tuned.

It is instructive to look at one example of a federal program in a specific field that has a proven record of achievement, state and local involvement, and political durability: the Agriculture Department's Extension Service, Cooperative State Research Service, and land-grant university system. The functions of identifying problems, planning research and development, evaluating new knowledge, and disseminating and applying it in the field, are all well integrated.

The first commercial hybrid seed was produced in 1923. One percent of the acreage of corn in the Corn Belt was planted with hybrid seed corn in 1933. More than 99% of the acreage of the Corn Belt and 92% of the nation's corn acreage were planted with hybrid seed in 1957, and nearly 100% in 1970.

7-2 PLAGUES AND INSECTICIDES

In the years 1348–1850, the Black Death, a pandemic plague, killed one-quarter to one-third of the population of Europe, was equally severe in Asia and Africa, and was transmitted to every corner of the world.

The disease was marked by the appearance of black blotches on the skin, caused by hemorrhages, and was transmitted, through fleas, from rats and other rodents. An agonizing death occured in 5–6 days. The outbreak appears to have originated in the Orient, and spread westward.

In 1347, the Mongol army of the Kipchak Khan Janibeg was riddled by the disease while besieging the Genoese port of Kaffa, in the Crimea. Janibeg catapulted the disease-ridden bodies of dead soliders into the town, thus spreading the disease among his enemies. The Black Death reached Sicily in the same year. The following year the disease had spread to North Africa, Italy, Corsica, Sardinia, Spain, and France. In August of 1348, a boat from Calais, France, carried the disease to Dorset, England. Early the next year, the plague had reached London, and, by 1350, every section of the British Isles had been infected. The disease recurred periodically and, by the year 1400, England's population of perhaps 4 million souls had dropped to about half that number. F. A. Gasquet, in 1893, published the first comprehensive study of the Black Death, *The Great Pestilence*.

In the early days of the Black Death there was an artificial and short-lived prosperity. The growing death rate had caused a massive redistribution of wealth. As the toll of human life continued, the economy began to respond to forces which were to prove catastrophic. The Dark Ages had returned.

The Black Death struck first in the cities and villages, where the population was most dense and the sanitary conditions most deplorable. There was no effective medical treatment for the plague and the only defense against infection was to flee from its presence. City-dwelling artisans were particularly hard hit. In England, an estimated 1000 towns and villages were depopulated. Inevitably there developed a severe shortage of skilled and semiskilled labor.

Many of the landowners who survived the Black Death fell to financial ruin. Since feudalism was based on a largely agrarian economy, the very basis of medieval society was undermined. Within a century, feudalism had crumbled in England, and elsewhere in Europe it was doomed.

Plague has been all but eradicated. There are fewer than 1000 plague fatalities per year, worldwide, and virtually none in Europe, where it once wrought such havoc.

The human struggle for survival, detailed in our history books over the ages, has been highlighted by tuberculosis, black death, plague, cholera, yellow fever, and malaria, all of which are endemic and epidemic.

The germ theory of disease was put forward by Pasteur little more than a hundred years ago. Physicians today fight diseases with drugs, insecticides, germidices, herbicides, and chemicals. Malaria is carried by mosquitoes. In India it is the monarch of diseases, with an annual death rate of one in every 100 cases. But because of the accompanying debilitation and suffering, the disease is associated with sociological effects far more widespread than death. DDT, invented in Switzerland by the Ciba Company, was first used to spray the insides of dwellings to kill mosquitoes, breaking the life cycle of the malaria parasite. In country after country, in the Far East, Africa, and refugee areas, the spraying program has brought about vast improvements in the health of large populations.

DDT is the poor man's chemical, at 17¢ per pound, and for 25 years has been the life saver of emerging peoples. Replacements, perhaps less persistent and more biodegradable, are far more costly. Nonetheless, there are serious arguments against the continued use of DDT because of its side effects; but is it not also proper to ask what are the consequences, beyond the preservation of certain species of birds, fish, and animals, in terms of human survival? Dr. Thomas H. Jukes, professor of medical physics at the University of California at Berkeley, feels that a ban on the use of DDT would violate the basic human right to be protected against deadly disease.

The World Health Organization (WHO), on January 22, 1971, stated that "the withdrawal of DDT would be a major tragedy in the chapter on human health. Vast populations in the malarious areas of the world would be condemned to the frightening ravages of endemic and epidemic malaria." The WHO has tested more than a thousand other insecticides and to date has not found substitutes for DDT. It has also examined the suggested hazards of DDT as a carcinogen and a mutagen and has rejected the alleged evidence.

More than 900 million people who were subject to endemic malaria are now free of it; 228 million live in areas where the disease is being attacked. Refugees from war, pestilence, natural disasters, floods, hurricanes, typhoons, and earthquakes are rid of lice and parasites, another triumph of DDT powder.

Thousands of potentially destructive insects invade the United States each year, threatening livestock, cereal, fruit crops, and plants. We spend a great deal of effort to prevent them from establishing themselves and inflicting terrible damage to our ecology and economy alike. Among the weapons developed by agriculturists are laboratory-bred sterilized flies (which effectively interrupt reproductive cycles), allure baits, colored traps, and black lights.

Over the years, winds, ships, planes, birds, mosquitoes, and travelers have brought to the United States more than 700 species of foreign-born pests. They include 50 plants, some as predatory as witchweed, whose tentacles suck the life from corn, sugarcane, and rice plants, and 55 strains of deadly diseases, some as contagious as "exotic newcastle," which once destroyed more than 11 million chickens in Southern California.

The female screwworm fly lays several hundred minute eggs in the slightest scratch or wound on livestock or wildlife, eggs that eventually hatch into flesh-eating larvae. Prior to 1962, they caused $125 million in yearly damage to the cattle industry in the Southwest. Even now, hordes of screwworm flies swarm up from Central America and Mexico each year. In a concrete structure in southern Texas, more than 100 people work to breed 26 million striped bluish-green screwworm flies per day. They funnel canisters of the fly pupae onto conveyor belts, where the developing insects are sterilized by gamma rays from cobalt-60. At dawn, a squadron of low-flying planes releases a steady stream of cartons, which pop open and rain millions of flies along the United States–Mexico border. The sterilized male flies, who outnumber the native males, win nearly all the females (who mate just once during their 3-week lifespans) and thereby block reproduction.

This type of biological warfare is one of the techniques of the U.S. Department of Agriculture's Animal and Plant Health Inspection Service (APHIS) at Hyattsville, Maryland. The service employs some 5300 scientists, veterinarians, inspectors, and assistants scattered across the country, from airports to swamps. They form a front line of defense against all species of invading insects and plants.

Their pilots soar aloft in small planes to study the migratory tendencies of airborne insects. Others ride horses through deserts and canyons to inspect any cattle or burros that stray across our 1900-mile-long border with Mexico, and thus keep out fever-carrying ticks. Other agents continually check battery-powered "black-light" traps mounted near international ports of entry. These lights produce a ray that attracts high-flying insects, which are then trapped in canisters.

When pests or contagious diseases sneak past peripheral defenses, quarantines must be carried out with precision. In a painstaking, grove-by-grove and block-by-block search in Texas' lower Rio Grande Valley, a squad of inspectors recently made certain that not a single back-yard tree contained the smallest speck of the sooty, mold-harboring eggs laid by the citris blackflies which had slipped in from Mexico in 1971.

In the summer of 1971, a Brownsville, Texas mare's sudden violent illness was confirmed by alert officials as a deadly sleeping sickness, Venezuelan equine encephalomyelitis (VEE). Carried from horse to horse, and to human beings by mosquitoes, the disease had been confined to northern South America until the late 1960s. By the time VEE reached Texas, over 10,000 horses and burros and 42 humans already lay dead in Mexico as its victims.

Quarantines were imposed, limiting the movements of horses from threatened areas. Forty-five planes sprayed a curtain of insecticide over a 540-mile-long stretch along the Rio Grande and Gulf Coast. Over 4000 veterinarians, plus federal, state, and county employees, were marshaled to help vaccinate every horse and pony in the entire Southwest with an experimental serum, the only known antidote for the disease. More than 2.8 million horses were vaccinated in 19 states from California to the Carolinas. Nonetheless, VEE struck 26 counties in southern Texas.

To destroy another type of enemy, five species of gnat-size black-and-yellow wasps have been turned loose in major dairy areas across the United States. These creatures ignore every insect except the alfalfa weevil that can so quickly ruin hay and pastures. Each species of wasp attacks weevils in a different age bracket.

The shiny green-and-copper Japanese beetle eats the fruit and flowers and skeletonizes the foliage of 275 different plants, while the infant grub beetle of the same species devours the roots. Scientists have recently discovered bacteria that retard the grub's development. Found to be harmless to all other forms of life, the bacteria were developed by the scientists into a beetle-killing powder which is now widely applied.

Pest fighters sometimes find it just as advantageous simply to confuse pests, and few are distracted as thoroughly as the elusive grayish-brown boll weevil. As the male weevil approaches any cotton field, a chemical scent, called a pheromone, lures him into a maze-like trap painted shades of yellow. The beetle travels through the maze, and once within it is unable to find his way out.

7–3   FORESTS AND INSECTS

In 1946, the Engelmann spruce forests in the higher Rocky Mountains of Colorado were a reservoir of unexploited virgin timber, the summer homes and playgrounds of thousands of people who love the mountains. Today, on much of that ground, stand millions of dead trees—graceless, lifeless, valueless. They will stand there 20 years more, ghost forests that are tragic evidence of how quickly and silently a tiny insect can do its damage when a combination of favorable factors has brought about a sudden increase in its numbers.

No person ever suspected what was happening until the outbreak was well under way and approaching its peak. It was then too late to do very much. Between 1942 and 1948, 4 billion board-feet of stumpage had been killed. The insects were more destructive than forest fires—over 6 years, 16 times more timber was destroyed than had been killed by fire in the previous 30 years in the Rocky Mountain region. The following parts of our National Forests were damaged for a generation: the White River, Grand Mesa, Routt, Arapaho, Uncompahgre, San Juan, and Dixie. In a large part of the White River National Forest nearly all the spruce of saw-timber size—2900 million board-feet—was killed.

In its adult stage the Engelmann spruce beetle is a small, cylindrical, hard-shelled beetle, about one-quarter inch in length and about the size of an ordinary housefly. When the adults leave the dead trees and start to fly in June and July they are reddish brown to black in color. They soon settle on recently felled or standing green trees and bore through the outer bark into the inner bark. This attack extends over most of the lower main stem of the tree.

The beetles work in male and female pairs, each pair raising a separate brood. The female makes the entrance, followed by the male, and bores a tunnel between the bark and wood which usually extends vertically and parallel to the grain of the wood. This tunnel is the "egg gallery." The eggs are laid in alternate groups along the sides of the egg gallery, which is packed with boring dust mixed with pitch. There are usually 3–4 groups of eggs and a total of about 125 eggs in each egg gallery. On the average there are 6–8 such egg galleries for each square foot of bark. When the eggs hatch, in 3–4 weeks, the larvae feed on the inner bark and cut mines that run at right angles to the egg gallery. This larval feeding continues through the late summer and fall. When winter arrives they are still in the inner bark, where they become dormant. The following spring the larvae resume feeding. As summer advances they become mature, transform to pupae, and then into adult beetles.

Previous outbreaks of the Engelmann spruce beetle occurred in the Rocky Mountain region, but most of them were long ago. In 1907, the Department of Agriculture found evidence on the White River National Forest of an outbreak that had occurred 20–25 years earlier. Severe outbreaks occurred in the Pike National Forest in about 1855 and in the Lincoln National Forest in New Mexico in about 1890. An outbreak that killed nearly 100% of the spruce volume swept over the Aquarius Plateau in Utah between 1918 and 1928. A localized outbreak was reported in the northwestern part of Yellowstone National Park in 1937. None of these earlier outbreaks approached the intensity and or total volume of destruction of the outbreak that started in Colorado in 1942. It was evident by 1943 that cutting must be immediate if this beetle-killed timber were to be used for lumber, since it would probably not remain usable for longer than 3–4 years after attack.

Direct-control measures, applied before infestation spreads over a large area, should prove effective. The insect can be destroyed by peeling and burning the infested bark, by burning the infested logs, or by the application of toxic, penetrating sprays. For example, a mixture of oil and orthodichlorobenzene has proved to be effective when applied to the bark of infested trees or logs.

The spruce budworm is a small, foliage-feeding caterpillar that periodically kills an immense amount of spruce and balsam fir in the Eastern States and Canada. It is a serious problem in jack pine forests in the Lake States, and in Douglas fir, alpine fir, white fir, Engelmann spruce, blue spruce, lodgepole pine, and ponderosa pine forests in the West. It is native to North America. Records of its ravages in the East date from about 1805. It appeared again in epidemic proportions in about 1880.

The first outbreak to be studied carefully began in Quebec in 1909, appeared in Maine in 1910, and in New Brunswick and Minnesota in 1913, continued for nearly a decade, and destroyed more than 250 million cords of spruce and fir pulpwood. About 30 million cords were killed in Maine, and more than 20 million in Minnesota.

An outbreak in Canada assumed epidemic proportions in 1935. By 1944, it was estimated that 125 million acres in Ontario were infested. In 1945, an official of a Canadian pulp and paper company claimed that the insect had killed enough timber to supply all Canadian pulp mills for 3 years. By 1947, most of the mature fir and a considerable part of the white spruce over an estimated 20,000 square miles had been killed, with less intense damage over a much larger area. The dead trees created a fire hazard; large areas affected by the budworm were burned.

The adult of the spruce budworm is a small moth with a wingspread of seven-eighths of an inch. Its general color is grayish with brown markings. In the Northeastern States the moths start emerging from their pupal cases around July 1. The females deposit their pale green eggs on the foliage in masses of 10–50 or more, where they overlap like fish scales. One female may lay several of these egg masses and on the average produces about 175 eggs. The incubation period lasts about 10 days.

After the eggs hatch, the young caterpillars crawl about until they find suitable places under bark or bud scales to spin silken, web-like coverings, or hibernacula, under which they spend the following fall and winter. These tiny larvae do not feed until they become active in late April or early May and leave their hibernacula. They are orange yellow at first and later turn brownish. They first mine the old needles first, and then enter the opening buds where they feed on the young needles which are just starting growth. They also feed on spruce and fir pollen. As the new shoots elongate, the larvae tie the needles together with silken threads and thus form shelters within which they feed. By late June they are full-grown and reddish brown in color, and begin forming the pupal cases, which are attached to the twigs. The pupal period lasts 7–10 days, after which the moths emerge and start laying eggs—a new generation is then under way.

Aerial applications of insecticides offer possibilities for controlling defoliators like the spruce budworm; attempts to control the insect over extensive areas by aerial spraying are now in progress. But all studies and observations by entomologists and foresters suggest that the ultimate solution lies in managing the forest in a way that maintains high vigor in balsam fir stands and, where conditions permit, increases the proportion of spruce.

Pine bark beetles are small, dark-colored, hard-shelled insects the size of a grain of rice or a medium-sized bean. They bore under the bark of various pines and dig egg tunnels, mostly in the inner bark, which cut the cambium layer—a tree's vital tissue. Eggs laid along the sides of these tunnels hatch into small, white, legless grubs. Under the bark, the attacking beetles also introduce fungi, blue stains, and yeasts, which penetrate the sapwood and plug the sap stream from roots to foliage. The tree is hurt in the same way that an animal would be injured or killed if worms were to bore into it and stop up all of its veins and arteries.

When the larvae complete their feeding in the inner bark, they change into pupae, the resting stage, and then to new adults. These adults later emerge from the bark and fly off to attack other pines. Thus they perpetuate their species and continue their destructive course. The new adults may attack the green trees nearby, or they may fly several miles to find trees to attack.

A great many different kinds of beetles work into and under the bark of pines. The most destructive bark beetles in American forests are the pine beetles, which attack primarily the more mature trees, and engraver beetles, which prefer young trees or the tops of older trees. Species of pine beetles and engraver beetles are found throughout North America.

The most important species of pine beetle are the western pine beetle, which attacks ponderosa pine and Coulter pine in the Pacific States, Idaho, Montana, and British Columbia; the southern pine beetle, which attacks all species of pines and spruce from Pennsylvania south to Florida and west to Arkansas and Texas; the mountain pine beetle, which attacks lodgepole pine, western white pine, sugar pine, and other pines in the Pacific States and northern Rocky Mountain regions; the Black Hills beetle, which attacks ponderosa and lodgepole pines in the southern and central Rocky Mountain regions and in the Black Hills of South Dakota; the Jeffrey pine beetle, which attacks Jeffrey pine in California; and the turpentine beetles, which attack all species of pines but can usually overcome only weakened and injured trees.

The engraver beetles attack all species of pines, breed readily in the tops of recently felled trees and in slash, usually develop large populations, and move into the tops of living pines, frequently killing trees in large groups. Various species are found in different parts of the country.

Because bark beetles are constantly at work in pine forests—thinning, harvesting, and wiping out entire stands of timber—they destroy on the whole a vast amount of commercially valuable timber.

Natural control factors keep some bark-beetle epidemics in check. Besides the limitations of food supply, disease, and unfavorable weather that restrict populations of bark beetles, they have a number of insect enemies. Parasites and predators feed upon and destroy the bark beetles. Also, many species of birds catch beetles when they are in flight. Certain species of woodpeckers go after beetle larvae which are in or under the bark.

Toxic oils sprayed on infested bark have been used to avoid the costs and fire hazards of peeling and burning it. Fuel-oil solutions of naphthalene, orthodichlorobenzene, and paradichlorobenzene have proved effective against the mountain pine beetle, the Black Hills beetle, and the engraver beetle in lodgepole pine and other thin-barked trees.

Early in 1946, the Douglas fir tussock moth appeared in epidemic proportions throughout a large forest area near Moscow, Idaho. A native of the northwestern United States and southeastern Canada, this small insect can kill its preferred hosts, Douglas fir and the true firs, in 1 year if it destroys all of their foliage; partial defoliation may result in serious top killing and the death of trees if it continues over several years.

In its life cycle this insect produces only one generation each year. Eggs are laid in August and September and hatch the following spring in late May.

The tiny caterpillars are active and will travel relatively long distances in search of food. They become full grown by late August, pupate, and transform to new adults in about 2 weeks. Since the female moths are wingless, eggs are usually laid on or near the pupal case from which the moths have emerged. Any widespread distribution of an infestation must be by means other than the flight of adult moths. It is known, however, that the young, hairy caterpillars can be carried long distances by air currents. When disturbed, they drop from the limbs and hang suspended on a fine silken thread often 5 ft or more in length, which they spin as they fall. This thread and the body hairs of the caterpillar offer considerable wind resistance, and air currents of about 10 mph will carry them away.

In 1947, it was estimated that within a gross area of about 500,000 acres, 350,000 acres with an estimated stand of 1,518,000 board-feet of Douglas fir and white fir timber were infested, and an additional 1,183,000 board-feet were threatened; the economic values involved (including, but not limited to, stumpage, lumber, payrolls, and taxes) were more than $100 million. Aerial spraying with a DDT solution was the only feasible means of control on the rugged terrain. The estimated cost of control was $1.70 per acre, and to hold down losses the operations would have to be accomplished between May 20 and June 30—May 20 was the approximate date of the general hatching of tussock moth caterpillars from the egg masses, and after June 30 the defoliation of trees would kill much of the timber stand.

In the field headquarters at the Moscow, Idaho, APHIS facility, a group of men waited for the signal that started the greatest of all airplane offensives against the tussock moth. When it came, airplane spraying spread 1000 gal of DDT spray over 1000 acres of infested fir timber. Spraying operations were eventually completed over a total of 413,469 acres of fir timberland, and 390,878 gal of spray were applied. No live tussock moth caterpillars were found a week after the spraying; the infestation had been stopped in its tracks. The cost of the project was just under $1.57 per acre, or about 13¢ less than the estimated cost.

Oregon's Blue Mountains now suffer the same infestation. In 1973, Douglas fir tussock moths caused visible defoliation in about 600,000 acres of timberland in eastern Oregon and in Washington. The total area infected reached about 900,000 acres, mostly on government land, and new outbreaks were spotted in British Columbia, Idaho, and Montana. "This is the worst infestation we have ever encountered," said Robert E. Dolph, an entomologist for the U.S. Forest Service in Portland. "We've seen a threefold increase in the acreage affected."

"We will salvage much of that timber," said forester Glenn Parsons, "but the growing stock has been lost on much of that land. It will take 50 years to correct the damage." Noting that tussock moth outbreaks are showing up in other states, Parsons added: "It looks like conditions are favorable for a spread throughout the West."

Occasionally the moth population explodes. The caterpillars, which sport long tufts of hair that give the species its name, chew their way through enough fir and pine needles to kill whole trees and weaken others. The Forest Service noted that two of Oregon's big forest fires involved moth-killed trees.

In the past, foresters used DDT to stop the tussock moth. But the EPA banned DDT, and when the Forest Service sought an exemption to cope with the expected outbreak, the EPA administrator refused. That decision enraged many private and government foresters, who warned that substitute pesticides would not work.

"It's discouraging to see the insect world gaining on us like this," said Secretary of Agriculture Earl L. Butz, adding his voice to the great tussock-moth controversy in what has been the largest effort yet to crack the Environmental Protection Agency's ban on DDT.

Tree diseases are of two main types, parasitic and nonparasitic. The parasitic, or infectious, diseases are frequently highly contagious. They are caused mainly by low forms of life, such as bacteria, fungi, viruses, microscopic eel worms, or nematodes, and by seed plants such as mistletoes and dodders.

Among the nonparasitic diseases are such disorders as sunscald, winter injury, drought injury, root drowning or suffocation, nutritional excesses and deficiencies, and injury from gases, smoke, and fumes.

The white pine blister rust offers an example of a disease that can be controlled through the removal of the alternate hosts—currants and gooseberries. The rust cannot spread directly from pine to pine, but spores from the rust on pine are carried by the wind and are able to infect currants and gooseberries. Spores from the rust developed on these plants are in turn capable of infecting white pines. The removal of currant and gooseberry bushes to a safe distance from white pines effectively protects the pines from the rust.

If a disease has become widespread and well established, eradication is usually impracticable, and we may have to learn to live with it and to reduce losses through indirect methods of control. This applies to most of our native diseases. The red rot of the ponderosa pine in the Western States is an example. The causal fungus enters the trunk through naturally occurring lower dead branches and eventually causes an average loss of about one-fourth of the total timber volume. It rarely enters through branches less than 1 inch in diameter, however. Control of the disease is possible by either pruning off the lower branches before they die or by growing the trees so close together that the lower branches are shaded out before they become large enough to support the fungus.

Varieties resistant to disease have been successful in field and fruit crops; there is every reason to expect that they will prove equally valuable in our future forest- and shade-tree plantings. Although work along this line has scarcely more than begun, an American elm resistant to the Dutch elm disease, other trees resistant to phloem necrosis, strains of mimosa resistant to the mimosa wilt, and white pine resistant to blister rust, have been tested. These resistant trees are now being propagated.

In September, 1973, Florida agricultural specialists began injecting palm trees with oxytetracycline to combat "lethal yellowing," a mycoplasmic disease that threatens to destroy most of Florida's 2 million coconut palms within the next 5 years. The program is designed to save the decorative palms, but it also could move oxytetracycline into plant disease control.

Florida officials are experimenting on some 10,000 trees. Three groups are being tested to determine the effects of oxytetracycline on lethal yellowing: diseased palms, healthy trees, and a third group as controls. The test will supply efficacy information as well as data on the feasibility of extended use. A variety of diseases are caused by mycoplasma or mycoplasma-like organisms. In California, the chemical is in experimental use on diseased pear trees.

The problems caused by the disease can be very serious indeed, and an International Council to Combat Lethal Yellowing has been formed by a group of plant pathologists doing research in several affected countries.

Some of our present-day kinds of trees, as indicated by fossil remains, flourished in North America millions of years ago; clearly, trees and their parasites must have fluctuated in abundance long before the coming of the human race. Then, as now, periodic epidemics must have caused extensive losses, but when the trees were attacked they could usually maintain themselves against borers and beetles.

The gypsy moth illustrates the serious consequences of the introduction of a forest insect from Europe. In 1869, a number of egg clusters of the gypsy moth were brought from France to Medford, Massachusetts, by a French mathematician and astronomer who hoped to develop a hardy silk-producing insect by crossing gypsy moths with silkworm moths. During his experiments some of the insects escaped. Some 20 years later the population of the gypsy moth had increased to a point where the damage was severe enough to attract general notice. At that time about 360 square miles were found to be infested. Within another 5 years, the infested area had increased to 2200 square miles. The gypsy moth, which defoliates both deciduous and evergreen trees, is now prevalent in New England, in an extensive area in eastern New York, and in an isolated area in Pennsylvania.

Chestnut blight has caused the complete destruction of our commercial chestnut, from Canada to the Gulf States. This record surpasses that of any other disease or insect. First reported in New York City in 1904, the disease spread rapidly. For many years roots of killed trees continue to send up sprouts, but these sprouts are usually killed by the disease before they are more than a few inches in diameter. Unfortunately, a 40-year search has not resulted in the discovery of a single American chestnut tree with sufficient resistance to be of practical value. Blight has reduced millions of acres of forest land to a lower productive status for an indefinite period because the native tree species replacing the chestnut are usually less valuable.

The smaller European elm bark beetle is an example of an introduced insect that was once of little importance until it became associated with the so-called Dutch elm disease fungus. The insect is known to have been established near Boston as early as 1904. It did little damage and was not considered a primary pest. In about 1930, however, when the Dutch elm disease fungus reached this country, the importance of the European elm bark beetle changed: it proved to be a carrier and transmitter of the fungus. The relationship works to the advantage of the bark beetle. American elms inoculated by contaminated beetles develop disease symptoms, are partially or completely killed by the disease, and provide suitable breeding material on which increasing populations of beetles can develop.

The elm beetle was unquestionably introduced through different ports. It and the fungus were present in burl-elm logs imported for veneer manufacture before quarantines had prohibited the movement of elm wood into this country. Beetles and larvae have been found in the elm wood used in certain types of crates received from Europe. A larger species of beetle, also a carrier of the Dutch elm disease fungus in Europe, similarly has been introduced into this country in burl-elm logs, but apparently has not been successful in establishing itself here.

We may expect that these two pests, and phloem necrosis, a virus disease, will eventually kill most of the elm forest growth in the northeastern quarter of the country.

The Dutch elm disease is caused by the fungus *Ceratostomella ulmi*. The disease was discovered in the Netherlands in 1919, and spread rapidly in Europe. It was first found in the United States in 1930; it too, had been brought here in elm logs imported for the veneer industry.

Native elms of the United States are dangerously susceptible to the fungus. Despite vigorous efforts to suppress it, the disease has become established in plantations and natural stands of the principal elm shade-tree areas of this country from Boston westward to Indiana and Kentucky and southward to Virginia. It has also been found in Tennessee, and an isolated outbreak was discovered in Colorado.

Dutch elm disease produces a wilting or yellowing of leaves on one or several branches, causing the leaves to fall. Later in the season, or in following years, the disease may spread to other parts of the tree until the entire top is affected and the tree dies. In more acute cases the entire tree may suddenly wilt and die with or without pronounced yellowing of the foliage.

Out of our experience with white pine blister rust, chestnut blight, and some forest insects that were known to have been imported from abroad, came the enactment, in 1912, of the Federal Plant Quarantine Act. The first quarantine prohibited further importation of white pines. The affected states also enacted laws to control the blister rust, or promulgated quarantines and regulations under established pest-control laws pertaining to the control of blister rust. Such action has been taken by 32 states. In 1917, a federal embargo was placed on the movement of white pines and ribes from the Eastern States to points west of the Great Plains to prevent westward extension of the disease through the shipment of infected host plants. This embargo was lifted in 1926, after it became evident that the disease had become widely scattered in western white pine forests. Adjustments were made from time to time in the federal white pine blister-rust quarantine to deal with problems created by the natural spread of the rust into uninfected territory and to effect the removal of ribes in control areas.

Control operations began in 1922 in cooperation with the Northeastern States. They were extended to the North Central, Southern Appalachian, Northwestern, and Pacific Coast regions as they were invaded by the disease, but 11 years elapsed before control work was well under way in all commercial white pine regions. One of the first steps in controlling the rust in each region was to delay its natural spread as much as possible by removing the cultivated European black currant. This plant is highly susceptible to rust and one of the chief agents in the long-distance spread and establishment of the disease in new locations. Its early removal was an important factor in retarding the advance of the disease.

The dwarf mistletoes are serious pests in western coniferous forests. The losses they inflict in terms of volume of timber and quality of lumber have never been accurately evaluated, but are believed to be exceeded only by the damage done by heart rots.

The dwarf mistletoes belong to the genus *Arceuthobium*, a group of the family *Loranthaceae*, of which all mistletoes and some other parasitic plants are members. Among their next of kin are the familiar Christmas mistletoes, which attack mostly deciduous trees and junipers. In North America, the junipers and their relatives are immune to dwarf mistletoes, although the generic name, *Arceuthobium*, is derived from Greek words meaning "juniper living," because juniper is their most common host in the Mediterranean region, where these plants were first described.

The economic importance of dwarf mistletoes is great. Damage by mistletoe in the forest is in four general categories: increased tree mortality, lower timber quality, reduced increment, and predisposition to other diseases or insect attack.

Mistletoe-infected trees are poor seed producers. Stands that are attacked by the parasite therefore do not reproduce as abundantly as do healthy ones. In addition, mistletoe retards the growth of its host tree. A number of studies in western states indicated that mistletoe may reduce the lumber production of a tree by 30–50%. The only method known for effective control of mistletoe is to prune it out and thereby eliminate the absorption system and the sources of reinfection. In the case of *A. vaginatum*, if an infected branch is cut 18 inches or more behind the mistletoe shoots the entire mistletoe plant is effectively removed from the tree.

Heart rots, which are caused by fungi that attack the wood of living trees, are to blame for an estimated annual loss of 1.5 billion board-feet in our commercial forests. The loss lies somewhere between the approximately $10 million value given the cull as stumpage and the $47 million value given it as logs. Every timber species in the United States is subject to attack by one or more species of the fungus, but fortunately a large part of the losses can be prevented by proper management.

When a fungus that is decaying the heartwood of a tree has developed for a number of years, it often produces a spore-bearing structure shaped like a mushroom or a bracket-shaped conk. Each year, one such structure can produce millions of tiny spores which are carried about by air currents. When a spore comes to rest on exposed wood and conditions are suitable, it germinates and sends fungus filaments into the wood. By means of these threads the fungus spreads through the tree, feeding upon and rotting the heartwood. Some of the fungi which cause our common root and butt decays rarely produce spores, but spread largely by growth through the soil.

Genetics has given us a good tool to use against the diseases and insects that attack trees—the selection and breeding of trees for resistance to pests. It is a long job. The time that a tree crop takes to produce seed and to mature exceeds the span of a human generation. Natural forces, aided now and then by man, have determined through the ages which forest species should servive, and these are the species with which the forester, the geneticist, and the forest pathologist now work.

The development of DDT during the war and of several more remarkable insecticides since then—among them, benzene hexachloride and chlordane—has presented an entirely new concept of the practicability of insecticidal control of forest insects. The toxicity of the arsenates or cryolite was so low that 15–30 lb per acre were needed to obtain the same degree of control as can be achieved with ½–1 lb of DDT. During the season of 1947, more than 500,000 acres of forest land were treated in various parts of the United States with DDT, at a rate of 1 lb per acre. Satisfactory control resulted at costs ranging from \$1–3 per acre. Much more remains to be done in perfecting equipment and improving the technique of application, but it seems safe to generalize that the control of defoliator outbreaks in the future will be a sound and economical forest operation. On the average, more than 2 million acres of forest land are defoliated annually. This entails the destruction of 10–75% of the trees in outbreaks of many different insects, and in all cases results in a tremendous growth reduction in the trees that are not killed outright. It does not appear too optimistic to hope that more than one-half of this loss can be prevented by aerial spraying with the chemical weapons now supplied by science.

7–4 FIRES

Perhaps, to the environmentalist, Smokey the Bear is an apostle worthy of affection; but let us look at *Trees*, the 1949 Yearbook of the Department of Agriculture.

According to this source, the main causes of forest fires are: smokers' matches or burning tobacco in any form; campfires; burning debris—fires which are originally set for clearing land or disposing of rubbish, stubble, etc., and which get beyond control; incendiary fires, which are deliberately started with the intention of burning over the land or damaging property owned by someone else; lightning; lumbering; and fires resulting from the maintenance of rights-of-way or from the construction or operation of railroads.

The worst forest fire in American history was the Peshtigo Fire in Wisconsin in October, 1871, when 1,280,000 acres were burned over; homes, towns, and settlements were swept away, and 1500 persons perished.

The principal causes of forest fires in the different sections are: In the Lake States and New England, fires are mostly man-caused; careless smokers head the list. In the Eastern and Southern states, also, they are mostly man-caused; careless brush and field burners and (in the South) intentional fire setters head the list. In the Rocky Mountain area, more than 70% are started by lightning; the others result from carelessness while smoking. In the Northwest, about half the fires start with lightning; careless smokers and campers are serious offenders. In California, lightning starts about 23%, careless smokers and campers most of the others.

The most destructive agents of our forests are forest insects and diseases, which account for more than double the losses each year by fire.

In the fall of 1970, fires burned all over Southern California for over a week. They charred 525,000 acres of brushland, destroyed 400 homes and 300 other buildings, and left 11 dead and 350 injured. In a state prone to the immoderate disasters of flooding, earthquake, and fire, it was the worst conflagration in history. Each acre of chaparral brush contains up to 30 tons of highly combustible fuel; the heat energy generated by the fires was equivalent to that of 12,500 Hiroshima bombs.

Natural conditions conspired at combustion. There had been no significant rain for 200 days, humidity was down to 5%, and temperatures climbed over 100°F. Hot, seasonal Santa Ana winds swept in from the desert to the northeast. Heavy rains during the two previous winters had nourished an unusually heavy undergrowth, now dust-dry.

The first ignition occurred in the Malibu area above Los Angeles. Sparked high in Las Virgenes Canyon, the flames spread across 50 acres in 5 min and were soon rushing toward the sea, consuming the houses of movie stars and businessmen. Another fire broke out in the north near Newhall, in the dry foothills of the Santa Susana Mountains. The two blazes later joined at the Ventura Freeway. Among the casualties were beach houses, homes, and ranches.

Many of those who had been burned out were determined to build again on the same sites, even though they knew that, by a perversity of nature, the fires, having burned off so much ground cover, were likely to bring on a flood disaster.

But besides naturally caused fires, others are being set by the tens of thousands each year by professional foresters who discovered that a controlled fire one year may prevent a catastrophic fire the following year. We may follow Smokey the Bear's advice, and should generally strive to prevent fires, but at the same time we are learning that wildfire is an integral part of the natural order.

The U.S. Forest Service, since 1905, has said "Don't let fires start—ever." John Muir noted the openness of California's Sierra Nevada forests, which seemed impervious to fire. In 1875, he had seen "cataracts of flame" in brush fields which became calm when they spread under the trees. Naturalists held that the Sierra forests had been periodically burned over by fires that had been started by lightning, which consumed brush and litter and eliminated the build-up of dangerous tinder.

Fire weather is a suitable combination of drought, heat, and wind that occurs annually in California and periodically in other temperate and tropical regions. Ignition occurs when people are careless or malicious, or when there is lightning; the latter, according to the Forest Service, causes half of the serious woodland fires on the West Coast.

In 1973, Northern and Central California had another very high acreage fire, as did the adjoining states of Nevada, Oregon, and their neighbors, Montana and Washington. Many thousands of trained fire fighters worked for weeks to calm the blazes, but the conflagration was greater than that of 1970.

Table 38. World's Worst Fires and Explosions

Date	Location	Deaths	Property Damage ($ millions)	Comments
9/2–6/1666	London, England	—	—	Fire destroyed 89 churches, 13,200 houses; 200,000 homeless
12/16/1835	New York, New York	—	20	About 700 buildings burned
12/8/1863	Santiago, Chile	2,000	—	Church of the Compania burned while filled with worshippers
10/8–14/1871	Michigan and Wisconsin	1,000	—	Forest fire devastated great areas of these states
3/10/1906	Courrieres, France	1,060	—	Mine explosion
12/6/1917	Halifax, N.S., Canada	1,500	35	20,000 homeless; explosion of war material
10/13–15/1918	Minnesota and Wisconsin	1,000	100	Forest fire
3/22/1934	Hakodate, Japan	1,500	—	Fire destroyed largest city north of Tokyo
11/12–16/1938	Changsha, China	2,000	—	Fire leveled city
6/8/1941	Smederevo, Yugoslavia	1,000	—	Most of town destroyed by explosion of ammunition plant
4/26/1942	Honkeiko Colliery, Japan	1,549	—	Worst mine disaster in history
9/4/1949	Chungking, China	1,700	—	Central part of city burned, 100,000 homeless
8/7/1956	Cali, Colombia	1,100	—	Seven trucks carrying dynamite exploded

For more than a generation, sportsmen have been buying abandoned cotton plantations in the Southeast, as well as cut-over timber lands, for hunting preserves. Quail favor the seeds and berries of the open lands. When the lands begin to grow thickets of young pine trees, there is no longer any food for the quail. Foresters have found that fire plays an essential role in reforestation. The loblolly, slash, and long-leaf pines reproduce poorly and require contact with bare soil in order to germinate. The sequoias of the West behave in a similar fashion. Fire can also control the fungus disease known as brown-needle spot. Forests that have been burned regularly "by prescription" now support more quail, deer, and other game.

Foresters learned to burn carefully, in small patches isolated by fire strips from which litter has been removed and to incorporate factors such as air temperature, humidity, wind velocity, and number of days since the last rain in their decisions. They have been successful over tens of thousands of acres in the Southeast. The Forest Service has now published "A Guide to Fire by Prescription."

Over much of the earth, fire seems to be a part of the natural order. Similar patterns of climate turn up the necessary weather conditions for fire every few years in the Carolinas and every summer in California. Fire cycles the mineral constituents of forest litter to provide nutrients. Cattle, grazed in burned-over forest or pasture, gain weight more rapidly than those grazed on unburned land. Catastrophic crown fires do indeed destroy wildlife, but these fires are the result of wasteful logging practices that do not involve fire.

Clearly, we should not stop trying to prevent forest fires, because prescribed fires are powerful tools that are dangerous in unskilled hands. In the broad-leaf woodlands of the American Northeast, there is no such thing as a "good" fire.

Even though great loss occurs in brush and forest fires, the world's most devastating fires and explosions have historically occurred in buildings, plants, or houses that are entirely built, controlled, and operated by people. The number of lives lost, as shown in Table 38, is miniscule as compared with the lives lost in earthquakes or floods.

CONSERVATION

8-1 THE NATIONAL PARKS AND MONUMENTS

Part of our humanization of the earth consists in our development of the various forms of conservation for areas whose habitability, according to modern standards, is questionable. These we set apart, provide access to, and improve by building temporary habitations for visitors. They comprise our heritage of national parks, national monuments, resort and recreation areas, local parks and preserves, nature trails, and national seashores.

The Grand Canyon is a spectacular and complex canyon system of the Colorado River, running from Lee's Ferry in Colorado to Grand Wash Cliffs in New Mexico, rich with its varying colors, oddly sculptured pinnacles, huge temples, and buttes. It is a profound experience to descend to the canyon floor and race down its winding, chaotic river in a frail-looking wooden dory, much as John Wesley Powell did on his first trip more than 100 years ago, when he commented on the total uselessness of the area. Along the way, you will meet the Havasupai Indians, who still make their home in the canyon; and you may learn about the formation of the canyon, the canyon rims, and their differing water-caused climates and ecology.

Baja California, a long, narrow peninsula reaching halfway down into Mexico, is a region of tumbled mountains, yawning chasms, desert plains, lonely shores, and barren islands. In this land of grandeur and subtle beauty are giant cacti; mud, sand, and mirages; the palm-shaded oases of Conon Quadalupe; the Gulf Coast; and Pacific lagoons and islands. Crawling, creeping, and jumping creatures inhabit the vast central desert, and in the sea are whales, sea lions, pelicans, and elephant seals.

Alaska brings you face to face with the "outrageous magnificence of an overwhelming land." Here is where you'll see scenic panoramas on a grand scale: glaciers as big as Rhode Island; the Brooks Range, which has been scaled by few; Mount McKinley, North America's highest mountain; dramatic weather, permafrost, and wild winds; volcanoes (e.g., Mt. Katmai National Monument); earthquakes; and unique plants and animals—Kenai moose, Dall sheep, and musk-ox.

"Thousands of tired, nerve-shaken, over-civilized people are beginning to find out that going to the mountains is going home; that wilderness is a necessity; and that mountain parks and reservations are useful, not only as fountains of timber and irrigating rivers, but as fountains of life."

This was written almost 75 years ago by the naturalist, John Muir. His words were recalled as Americans celebrated the 100th anniversary of the birth of a great idea: national parks.

It was in March, 1872, that President Grant signed into law an act of Congress creating Yellowstone National Park. It was the first national park in the world. In 1864, President Lincoln had signed a law setting aside Yosemite Valley and Mariposa Grove "to be held for public use . . . for all time."

The idea of national parks grew slowly at first. Yosemite became a national park in 1890, as did Sequoia and General Grant national parks—the latter now being part of Kings Canyon National Park. Today the national park system includes 284 units embracing some of the nation's most magnificent scenery as well as landmarks of cultural and historical significance.

The centennial observations were not merely self-congratulatory. With problems of over-use in some areas, with our need to combine preservation with maximum enjoyment by everyone, with the goal of bringing parks to the people, the National Park Service scheduled various conferences to explore new directions in conservation.

Typical of these new directions is the recent proposal for a 24,000-acre Golden Gate National Recreation Area. Readily accessible to city dwellers, the park would include 30 miles of shoreline north and south of the Golden Gate, as well as Angel and Alcatraz islands in San Francisco Bay.

A wide variety of events heralded the flowering of a great idea that was born in 1872, more than 100 years ago.

Bordered by a freeway and a busy highway and overlooking the subdivisions and industries of Martinez is a pleasant knoll on which sits a shrine to the nation's foremost conservationist: John Muir.

John Muir had a passion for the wilderness, and his growing interest in conservation projected onto the national consciousness a previously unarticulated feeling for the need to halt the wasting of our natural resources.

His first book, *The Mountains of California*, appeared in 1894. It touched off a successful movement to preserve the nation's forests. His writings were instrumental in creating the U.S. Forest Service; Yosemite, Sequoia, and Mt. Rainier National Parks; the Petrified Forest National Monument; and part of what is now Grand Canyon National Park.

He was on personal terms with five presidents and did not hesitate to expound his conservation views to them. When Theodore Roosevelt toured

Yosemite, in 1903, Muir held long talks with him, and during the remainder of Roosevelt's term more than 148 million acres of national forests, 23 national monuments, and five national parks were added to the nation's park system.

The first explorers ever to peer into the depths of Yosemite Valley were probably members of Joseph Walker's exploration party, who, in 1833, were attempting the initial crossing of the Sierra Nevada from east to west. However, they made little note of the chasm since to them it was merely another obstacle in their struggle to traverse the mountains.

Then gold was discovered in 1848. Miners began to flock to the Sierra foothills. William Abrams and U. N. Reamer apparently entered the valley in 1849, looking for a sawmill site. But their expedition did not become known until Abram's diary came to light a century later.

It was Major J. D. Savage who led a band of the Mariposa Battalion into the canyon in search of Indians in March, 1851. Their reports touched off a flurry of interest, and, in 1855, James Mason Hutchings and two Indians guided the first tourist party into Yosemite. His published accounts soon brought visitors from all over the world.

Meanwhile the original occupants of the valley were falling on hard times. Chief Tenaya and his tribe, removed by Savage and his battalion to a reservation near Fresno, were permitted to return. But another "war" erupted between settlers and Indians, and the Army sent a force into Yosemite. Tenaya and his braves fled eastward to live with the Monos. One legend has it that they eventually stole some horses from their hosts and returned in a roundabout way to Yosemite. There, the story goes, they were surprised by a vengeful Mono war party and the old chief and most of his men were slain.

In 1864, moved by fears for the valley's future, Congress passed, and President Lincoln signed into law, a bill preserving Yosemite Valley and Mariposa Grove "to be held for public use, resort, and recreation, unalienable for all time." The land was ceded to the state for a park.

Four years later, John Muir stepped off a ship in San Francisco and set off on foot for Yosemite. The region cast such a spell upon him that he devoted much of the rest of his life to writing about Yosemite and the Sierra and working to preserve their beauties. He was instrumental in securing the creation of Yosemite National Park, in 1890. The park embraced the spectacular high country around the valley and became a unified operation in 1906, when California re-ceded the valley and Big Trees to the federal government.

At first, visitors could get to the valley only on foot or horseback, but in 1874 and 1875 two toll roads were completed into Yosemite. Regular stages began running in 1877, and hotels were built to accommodate the many visitors. The first auto successfully, but illegally, navigated the tortuous dirt roads into Yosemite Valley in 1900, but it was not until 1914 that the first motor-stage line was inaugurated. Beginning in 1907, visitors also could get to the valley by taking a train to El Portal and then transferring to stagecoaches (later replaced by buses). In 1926, a highway from Merced was completed, replacing the switchback Big Oak Flat road and the Wawona road as the main entry into the park.

Despite the occasional crowding, the scenery in most areas of Yosemite is as grand as ever. Congress, in 1913, permitted the building of a dam that flooded the Hetch-Hetchy Canyon on the Tuolumne to provide a water supply for the city of San Francisco, but elsewhere the peaks and meadows offer scenes of splendor.

In the valley is El Capitan, a massive rock twice as large as Gibraltar that soars almost vertically more than 3500 ft above the valley floor. Half Dome, a veritable symbol of Yosemite, stands 4748 ft above the valley, a massive testimonial to the grinding force of the glaciers which carried away half of the original mountain. The Mariposa Grove of giant sequoias is Yosemite's largest grove of big trees.

Tuolumne Meadows, at 8600 ft on the Tioga Pass highway, is the largest sub-Alpine meadow in the Sierra. From there, hiking and packing trails branch in all directions, reaching high into the surrounding peaks, some of which are snow-capped all year around.

In few other places in the world is so much grandeur compressed into so small an area as in Yosemite Valley. The sight of the grassy valley floor ringed by granite cliffs 4000 ft high or more, of cataracts that tumble over the walls, of lacy falls that seem to vanish into thin air, of shifting lights and shadows, gives a magical quality to the scene.

And yet the valley, about a mile wide and 7 miles long, is only a fraction of the 1189 square miles of Sierra wilderness that makes up Yosemite National Park. In the high country surrounding the canyon are great bald domes of granite, jagged peaks reflected in icy lakes, pleasant meadows and groves of giant redwoods.

Through it all, valley and High Sierra, wind more than 700 miles of trails and 200 miles of roads. Approximately 2.3 million visitors—rock climbers, back packers, fishermen, campers, motorists, and cyclists—travel there annually.

To accommodate these crowds and prevent the natural beauty of the park from being trampled to extinction, the National Park Service has adopted a number of innovative policies. The use of private vehicles has been discouraged by the establishment of open-air buses in the valley and at the Mariposa Grove of Big Trees. The attractions of the many close-in camping areas in the high country have been stressed.

An area about the size of Rhode Island is contained within the park boundaries. The park includes, in the words of Muir,

> the headwaters of the Tuolumne and Merced, two of the most songful rivers in the world; innumerable lakes and waterfalls, and smooth, silky lawns; the noble forests,

the brightest crystalline pavements, and showy mountains soaring into the sky 12,000–13,000 feet, arrayed in open ranks and spiry pinnacled groups partially separated by tremendous canyons and amphitheaters; gardens on their sunny brows, avalanches thundering down their long white slopes.

It is a land sculptured by the relentless hand of nature. Eons ago, its once-rolling hills were thrust upward and deep V-shaped canyons were cut by fast-running rivers. Then great glaciers moved down these canyons, squaring off their sides and leaving, as they retreated, spectacular polished vertical cliffs like those that enclose Yosemite Valley. The streams that once flowed into the valley were left hanging thousands of feet above, so that they now plunge downward in cataracts to join the crystalline waters of the Merced.

One hundred million years ago, when dinosaurs roamed the earth and strange winged reptiles were emerging from primordial swamps, there were already standing in stately groves around the world the ancestors of California's giant redwoods. The huge Sierra redwoods can reach ages of 4000 years, while some closely related Coast Redwoods (e.g., *Sequoia sempervirens*) date back at least 2000 years. Both varieties descend from a common ancestor, possibly the prehistoric "dawn Redwoods," some of which have been found still growing in a remote province of central China.

Limited to a narrow belt along the coastline of Northern California and southern Oregon, the Coast Redwoods have been preserved in numerous parks. Largest of these is Redwood National Park, a narrow, 58,000-acre preserve of beaches and forests along the foggy and cliff-bound coast north of Eureka, California. Included in its boundaries are three state parks: Jedediah Smith, Del Norte, and Prairie Creek.

Yurok and Pomo Indians who lived in the region generally avoided the dark redwood forests, where game was scarce and passage often difficult. The first foreigners to record the Coast Redwoods were members of the Portola expedition, searching for Monterey, in 1769. Jedediah Smith struggled through the rain forest in the general area of the park that now bears his name.

With the discovery of gold and the resulting great migration to California, the demand for timber soared. At first the redwood trees were spared; with only hand axes and short saws it sometimes took a two-man chopping team one week to fell a redwood, and even then the ox teams might not be able to haul the fallen giant away. But with the advent of two-man saws up to 16 feet long, steam-driven "donkey engines" to winch out the logs, and the building of larger mills, whole forests were cut down in logging. Fleets of small steam schooners loaded lumber along the North Coast, the cargo swung aboard by aerial cables.

Even as the logging started, efforts to spare the redwoods began. The most significant step was the formation, in 1918, of the Save the Redwoods League. Since its inception, the League has raised more than $16 million in donations and matching state funds to add more than 120,000 acres to the state park system, and more than 280 memorial groves have been created through donations to the League.

The redwood lumber industries, adopting conservation principles and moving toward production on a sustained-yield basis, have also contributed some groves of redwoods to the park system. An unusual aspect of Redwood National Park is that privately owned forested watersheds lying outside the present park boundaries have been proposed for cooperative development compatible with park operation.

Prairie Creek State Park, created in 1923, is a forest that includes the Gold Bluffs area, where gold dust is still mingled with the dark beach sand. In park meadows browse wild herds of elk. Under the giant redwoods and Douglas firs is an undergrowth of ferns, rhododendrons, azaleas, vines, and a variety of shade-loving wild flowers.

Del Norte includes many miles of beach, with tidal pools and crashing surf. Jedediah Smith features Stout Memorial Grove, an area of redwoods, and the Howland Hill Road, which provides a rewarding drive through the redwood region. Not part of Redwood National Park are Richardson Grove and Humbolt, the largest of California's state redwood parks.

Before California was settled, coast redwoods covered about 2 million acres. Of the approximately 1½ million acres remaining, about 235,000 are virgin timber. Although less than half of this is in public parks, conservation efforts are continually adding to this total.

The drive up the coast on Route 1 from San Francisco to the Redwood National Park takes the tourist through the Golden Gate National Recreation area, including Alcatraz Island, into the Point Reyes National Seashore, and then along some of the coast's most spectacular rocky headlands to historic Fort Bragg. From Fort Bragg it is only a short swing into a series of state redwood parks and along the Avenue of the Giants, which starts 50 miles south of Eureka and leads eventually to the Redwood National Park, one of the newest of our national parks.

The redwoods themselves—this primeval forest of *Sequoia sempervirens* which has seen as much history as scores of human generations—seems to inspire in man a reverence for life and simple awe from being in the company of living things that may survive for as long as 2000 years. The 45-mile-long park is strung along the California coast midway between San Francisco and Portland, Oregon.

A crisis was reached in 1964, when the National Park Service made known the results of a survey of the redwoods region which revealed that of the original forest of 2 million acres only 15% remained. If lumbering were continued at the then-current rate, virtually all old-growth redwoods not protected in the state parks—only 2% of the original number—would vanish at some point between 1984 and 1994.

The North Coast relies to a great extent on lumbering for its economy. Three state parks have already tied up 27,468 acres of prime virgin growth, and the logging lobby is convinced that this number is about 20,000 acres more than necessary.

Mrs. Lyndon Johnson lobbied on behalf of the California conservationists in Washington, while the boundaries of the park were being hammered out at hearings and in committee sessions. And it was Mrs. Johnson who dedicated the park, in November, 1968, after the battle concluded with a compromise spurred by a wave of national concern over the deterioration of the environment. The park now encompasses 56,200 acres.

Stream-bed diversion threatened Tall Trees Grove, a narrow jut on Redwood Creek which houses the tallest of all the redwoods. For all their majestic size, the redwoods are shallow rooted and maintain only a tenuous grip on the yielding forest floor.

Only a few minutes ride from almost anywhere in the metropolis that is East Bay are unspoiled wilderness areas, silent groves of stately redwoods, grassy hilltops with magnificent vistas, and sandy beaches where the cries of gulls supplant the roar of traffic. These are parts of the far-flung domain of the East Bay Regional Park District. The district operates some 27 parks on more than 29,000 acres of land within Alameda and Contra Costa counties.

The heart of the park system is the vast, tree-canopied crest of the hills which rise behind Berkeley, Oakland, and San Leandro. Here are the giant regional parks: Wildcat Canyon, Tilden, Robert Sibley, Redwood, and Chabot. But the district also includes more than a score of other parks, ranging from a bleak island in the bay to a 5-mile-long lake enfolded in the Livermore hills. This park system was born in 1934.

Shadow Cliffs was once a bleak and depleted sand and gravel pit near Pleasanton. Donated to the district by Kaiser Industries, it is now a well-populated swimming hole. On the shore of San Francisco Bay is the 2-mile-long Alameda Memorial Beach, a big sandy cove where once stood an amusement park, Neptune Beach. Point Pinole Regional Park, a secluded and undisturbed sweep of shoreline with a spectacular view of the bay, is a new acquisition. It encompasses high bluffs, topped by eucalyptus trees, rolling meadows, marshlands, and ancient bunkers used in the manufacture and storage of explosives.

A "mini" mountain range forms much of Coyote Hills in southern Alameda County. Located at the edge of the bay, the park abounds with rabbits, gray fox, raccoons, and wildfowl. Here are ancient Indian shellmounds—relics of villages 4000 years old—and a biosonar laboratory where scientists "converse" with seals and sea lions through sonic waves.

Rolling hills of grass, dotted with splendid specimens of valley oak, make up the 3100-acre Briones Regional Park that lies between Lafayette and Martinez. Part of the Mexican land grant, the area has changed little from the days of Felipe Briones who settled there in 1829, and who was killed in 1840, when he gave chase to Indians who had run off some of his cattle.

Another urban wilderness is Las Trampas Ridge, a regional park accessible by foot or horseback. It takes its name from the Indians' practice of hunting elk by driving them into box canyons with walls so steep that they formed natural traps ("las trampas"). The area remains in its primitive condition and still abounds with deer, bobcats, and other animals—including an occasional mountain lion. The park, which lies west of Danville and San Ramon, bars motor vehicles beyond the parking lot.

The list of other parks is long: from Sunol Valley, another large urban wilderness area of spectacular beauty, to the modest Contra Loma Park, a small reservoir for boating and swimming developed not far from the industrial waterfront of Antioch. An area of old coal-mining ghost towns has also been acquired for park purposes.

Burney Falls thunders over a 129-ft-high vertical cliff into the verdant depths of a narrow canyon. The falls are the outlet for a vast underground reservoir. They flow with a remarkable evenness through winter and summer, drought and deluge. The falls are unusual not just because of the torrent of white water that plunges over the cliff, but also because of the many small streams which issue out of the porous, fern- and moss-covered rock itself.

Situated in the heart of the Pit River country of northeastern California, the park includes 565 acres of forestland and a section of shoreline on Lake Britton, a 9-mile-long man-made lake that is part of the Pit River hydroelectric system. The park is midway between Mt. Lassen and Mt. Shasta, in an area reflecting great volcanic activity. Lava flows covered and re-covered the land, but rain easily penetrates this porous rock, and many ancient rivers and streams continued to run beneath rocky surface, some of them eventually swelling from the ground as full-fledged rivers.

Burney Creek rises from the earth only a short distance above the falls. Its waters flow into Lake Britton and eventually into the Pit River. At one time the creek tumbled directly into the river, but the falls gradually moved upstream as water ate into the sedimentary deposits under the lava cap, which broke off in chunks as it was undercut. This process is still going on.

An old Indian trail ran through the park, a trail followed by Hudson's Bay Company fur trappers as early as 1820. They found that the Indians dug pits 10–12 ft deep, placed deer antlers on the bottom, and covered them with brush in order to trap game. It was from this practice that the Pit River got its name. The abundant water power of the region, typified by Burney Falls and the Pit River, brought suggestions that the area could be a "Pittsburgh of the Pacific." But today Burney Valley and the rest of the Pit River area is still a sparsely settled, little-known region. Hydroelectric plants transform some of the power of the falling water into electricity. McArthur–Burney Falls Memorial State Park came into being in 1920.

From Monterey south to San Luis Obispo, State Highway 1 winds through some of the most spectacular scenery in California. For most of its 135 miles, the road follows the coastline, where the Santa Lucia Mountains crowd the sea. On one hand are fog-shrouded redwood forests and towering peaks; on the other is the ocean, where migrating gray whales still spout in the distance and sea lions colonize the offshore rocks. At the southern end of this stretch of coast, the terrain and climate are very hospitable and various peoples have lived there for more than 9000 years. No cities were built, no armies were raised, and the inhabitants instead learned how to fish the sea and harvest nature's bounty on land.

Morro Rock, towering 576 ft above the surf, may have been named by Juan Rodriguez Cabrillo in his voyage of discovery up the California coast in 1542. Soon the rocky dome, "El Moro," was used as a landfall by navigators on Spanish galleons that sailed the great circle route from Manila to Acapulco. The captain of one of these vessels, Pedro de Unamuno, put into Morro Bay in 1587, claiming the land for Spain. And 182 years later, in 1769, Don Gaspar de Portola and his party camped near Morro Rock on their march up the coast to find Monterey, noting that the rock was an island at high tide, "a little less than a gunshot" from shore.

The padres and settlers who followed in the path of Portola found that this coastal area of what is now San Luis Obispo County was a bountiful land. When starvation threatened settlers in the Monterey area 3 years later, enough grizzly bears were killed in Canade Los Osos, near the present Montana de Oro State Park, to feed the settlers for 3 months. Clams were so numerous on Pismo Beach that in later years farmers used to plow the sand and scoop up the clams to feed their hogs.

During Mexico's rule of California great grants of land were made, and cattle and dairy "ranchos" covered the rolling hills beside the ocean. Much of this land was later acquired by Americans and primitive seaport towns sprang up at Morro, Avila, Port Harford, and Pismo. In the late 1800s, quarrying was started on the sides of Morro Rock to provide material for breakwaters—an action that changed the shape of the monolith and eventually led to the creation of Morro Bay State Park and the saving of the rock from total destruction.

Today there are seven state beaches and parks along the relatively short stretch of coastline from Estero Bay to Pismo Beach. In the north are Cayucos State Beach and Atascadero State Beach, the only one of the three that offers overnight camping. Morro Bay State Park is at the heart of this chain of parks. It encompasses Morro Rock and the great sand spit that forms the seaward side of Morro Harbor, as well as a large forested area stretching from the harbor to the top of Black Hill.

Just to the south is Montana de Oro State Park, a 6800-acre preserve that encompasses much of the wild headland of Point Buchon, sometimes called "the Pecho Coast" after the Los Osos and Pecho y Islay ranchos. Relatively undeveloped, with primitive campsites, Montana de Oro has been changed little by time. The grizzlies are gone, but deer and other wildlife abound. One can visualize the days when settlers smuggled hides to waiting Yankee ships in defiance of Mexican laws, or the smuggling in of contraband whiskey during prohibition.

The south end of this chain of parks consists of Avila State Beach and Pismo State Beach, where clamming is permitted, but only under strict regulation, since biologists are working to replenish these once-plentiful beds.

The Point Reyes Peninsula is truly another world, only minutes from the metropolitan communities of the San Francisco Bay Area. This great triangular promontory on the Marin County coastline is a land that time forgot, a geologic island sliding up the coast along the San Andreas Fault at an average rate of 2 inches per year. It is separated from the mainland by the long and narrow Tomales Bay to the north and Bolinas Lagoon to the south. Its pasturelands and forested hills, its long beaches and sheltered cover, have changed little with the passage of time. Today, most of this peninsula is preserved in the Point Reyes National Seashore, a 64,000-acre park authorized by Congress in 1962.

The remains of more than 100 Indian Villages have been found on the peninsula, and it was probably more heavily populated centuries ago than it is today. When Mission San Rafael was founded, in 1817, many of the Indians on the point were drawn to the mission by the prospect of better living conditions. Some of the remaining Indians became *vaqueros* (cowboys) on the cattle and dairy ranches there.

The park's 10-mile-long beach on the western shore is spectacular, but undertows make it lethal for swimming. The sheltered Drakes Cove, however, has a good swimming beach, and nature trails crisscross the mountain range that rises behind park headquarters.

Meandering through the city of Chico on its way to the Sacramento River, Big Chico Creek is shaded by giant oaks and sycamores. The stream and the lush green belt that borders it form the heart of Bidwell Park—the third largest municipal park in the nation. A legacy from one of California's pioneers, Bidwell Park is more than twice as large as San Francisco's Golden Gate Park. It extends for more than 10 miles, from the valley floor to deep into Iron Canyon, a cleft in the foothills of the Sierra Nevada.

Farther upstream is Five Mile Pool—an even larger lake—as well as a variety of old-fashioned swimming holes in the lava-walled canyon at the upper end of the park. Hooker Oak is more than 1000 years old, its great gnarled branches propped up by concrete pylons, and is the largest valley oak known. The upper portion of Bidwell Park takes in both the lower reaches of Iron Canyon and the edge of the mesa that rises abruptly along the canyon's northern side.

The two smallest of the Channel Islands off Southern California are a national monument. Anacapa and Santa Barbara islands, like their neighbors, were once part of mainland mountain chains. A great subsidence occurred and the mountains sank beneath the sea until only the tips of eight peaks protruded. Descendents of many of the plants and animals that lived on those ancient peaks survive on the islands today, often in forms quite different than their mainland counterparts.

Of the five northern Channel Islands, only Anacapa and Santa Barbara are open to the public. The two largest islands, Santa Cruz and Santa Rosa, are cattle ranches; San Miguel is a Navy preserve. On the rocks and beaches and in the grottos of these islands live the greatest variety of seals and sea lions found outside the polar regions. On San Miguel are seals and sea lions in numbers estimated at 12,000–20,000. Great sea elephants, once almost extinct, number about 3500. The Alaska fur seal from the Pribiloff Islands also breeds here.

Anacapa, about 10 miles from the mainland, is a chain about 5 miles long, of three small islands. Ancient lava flows festoon the vertical cliffs that ring the islands and wave action has eroded the porous rock into caves, grottos, and arches. Tide pools shelter varieties of small marine life. Santa Barbara, smaller and more remote than Anacapa, is also circled by high cliffs.

The Channel Islands were visited by Juan Rodriguez Cabrillo in 1542. Sailing up the California coast in search of the Northwest Passage, Cabrillo put in at San Miguel to wait out the winter storms. He died there in 1543.

The Canalino Indians, a tribe of the Chumash, lived on the island for many years. They plied the channel in big 12-man whaleboats made of planks tied together with leather thongs and caulked with bitumen. In the late eighteenth century, Russian seal hunters with crews of Kodiak Indians drove off the natives while slaughtering the island's seal and otter population. Since then the Channel Islands have been lightly populated. Santa Barbara, Anacapa, and San Miguel have no one living on them. The lighthouse on Anacapa is automated.

8–2 AFRICAN GAME LANDS

Although the National Park System in the United States is more than 100 years old and is mandated for multiple land use on a national basis, the parks and preserves in the savannah zones of Africa have come into being only in the last quarter-century.

The parks in Kenya, Tanzania, Uganda, and Zambia total 38,000 square miles (Figure 28), an area the size of New England. The various parks in Malawi, Botswana, Rwanda, eastern Zaire, and Ethiopia, together with a string of similar areas in Mozambique, Rhodesia, and South Africa, make up almost 40,000 square miles. There are similar parks along the western side of the continent.

Most of these are savannah parks, but others are mountain or marine areas. Tsavo and Kafue are both over 8000 square miles, Kruger is over 7000;

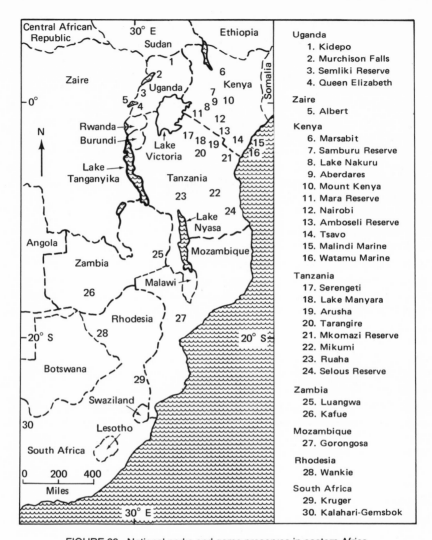

FIGURE 28. National parks and game preserves in eastern Africa.

Serengetti, 6000; Wankie, 5500; Luangwa Valley, 5000; Ruaha, 4400; Kalahari–Gemsbok, 3600; and Albert (once Kini, now Virunga), 3000. Yellowstone is 3400 square miles, the largest national park in the United States.

Meyers[1] holds that the parks are too small because of human competition for the land and for the protein food the animals could provide, and because of

[1]N. MEYERS, *Science 178:* 1255–1263 (1972).

the ever-increasing agricultural needs that parallel population growth. Serengetti has 1.5 million herbivores, some of which have migrated 24 miles outside of the park. Kruger has so many elephants that they are destroying their food supply faster than it can grow, so that excess elephants are now being cropped. Given the limited agricultural potential of Africa, there is a greater population density than is recognized.[2] And instead of finding a strategic retreat in the hinterlands for their migrations, wild creatures are encountering competition for water and for grazing and living room.

The 15,000 square miles of Kenya's Masailand, hitherto regarded as tribal property and having little hindrance to wild animals, have recently been completely demarcated and accorded to individual and group ownership. This entails the introduction of fences and modern ranching techniques, which clearly disturb the migratory patterns of wild animals.

Zambia still has plenty of room. Botswana has expanded its parks to one-fifth of its total territory. However, every district surrounding the chain of wildlife areas in Tanzania—Serengetti, Ngorongoro, Manyara, Tarangire, and Arusha—is already classified as being at full capacity or overpopulated. Kruger has tackled the problem of excess animal populations by converting to a sanctuary, and has a 500-mile-perimeter fence built to withstand elephants.

Tourism is the principal use to which the parks and reserves are put. Kenya had 400,000 visitors in 1971, projected to increase 20%. With tourism and foreign exchange, Kenya is more of a tourist country than Italy or Mexico.

The merits of game cropping have been discussed in detail.[3,4] They are matched by the economic potential of harvesting wild creatures. Murchison Falls, Queen Elizabeth, Mkomazi, Luangwa Valley, Wankie, and Kruger parks have adopted cropping as emergency measures. Cropping can be very remunerative to park funds. At Ngorongoro there is multiple land utilization, involving not only the fostering of wildlife activities, but also cultivation, animal husbandry (pastoralism), and forestry and watershed management. Cropping is practiced, and is seen as a harvesting earth's bounty. The Ngorongoro crater is safeguarded. The water tables of the 100-square-mile crater, dependent on catchment areas 20 miles beyond its rim, are carefully controlled.

Several other parks are employing multiple land use. Queen Elizabeth Park in Uganda engages in massive management of its hippopotami, and in prior years has served as a sanctuary for elephants, whose population doubled during the 1960s. It encourages tourism and applies modern techniques of forestry. Commercial fishermen operate within its boundaries, and meat from

[2] W. A. ALLAN, *The African Husbandman*, Oliver & Boyd, Edinburg, 1965.

[3] L. M. TALBOT, W. J. A. PAYNE, H. P. LEDGER, L. D. VERDCOURT, and M. H. TALBOTT, The meat production potential of wild animals in Africa: A review of biological knowledge. *Commonwealth Agricultural Bureau, Tech. Commun.* No. 16, Farnham Royal, U.K., (1965).

[4] J. A. BINDERNAGEL, *Game Cropping in Uganda*, Uganda Game Department, Entebbe, 1968.

its wild animals is supplied as food to local communities. It is expected that
other African parks will soon follow its pioneering example.

8-3   THE LAKES AND PARKS OF CALIFORNIA

Civilized societies often convert nature's waste into national parks or
monuments. Stories of "burning mountains," their rocky peaks glowing with
fire and smoke, were told by early gold seekers who had taken the northern
trails into California. This was the land made famous by Peter Lassen, the
Danish blacksmith and trailblazer. It was, and is, an area of steaming
fumaroles and bubbling lakes, of lush meadows close by the rocky ruins of
shattered pinnacles, of volcanic craters rising amid great forests.

It is a land of lakes. There are at least 100 of them in Lassen Volcanic
National Park alone, and about 40 of these are of respectable size. Many are
cold and blue, some are green and hot. The visitor to Lassen Park is naturally
drawn to Manzanita and Reflection lakes at the west entrance. Cold Boiling
Lake bubbles not from heat but from carbon dioxide gases seeping up through
it, while its shores are alive with bubbling mud pots and steam vents. Boiling
Lake, on the other hand, steams at an average temperature of 125°F. Soda
Lake is filled with naturally carbonated water, and Snag Lake is the result of a
lava flow cutting off a stream less than 200 years ago. Trails crisscross the
park, and the visitor can easily step into another world where wildlife abounds
and myriad lakes reflect the now-silent majesty of Lassen Peak.

Although legend has said little about the volcanic activity here, it was an
area of violent and recent eruptions. The southernmost peak in the volcanically
formed Cascade Range, Lassen Peak—like Mauna Loa in Hawaii—is part of
the "Rim of Fire," the great circle of volcanoes which rings the Pacific
Ocean.

On May 30, 1914, at about 5 P.M., Lassen again became an active
volcano. A great puff of steam, gases, and ash erupted from a new crater on
the north slope. During the next few months, eruptions continued intermit-
tently, and pictures of the towering volcanic clouds made front-page news
around the world.

There was a series of 150 eruptions of gas, steam, and cinders. Then, on
May 19, 1915, molten lava bubbled over the rim, melting snow and forming a
great flood of mud that washed down the valleys of Hat Creek and Lost Creek
on the northeast slope. The lava formed a lid on the volcano, and for three days
pressures inside built up until there was a terrific blowout from the side of the
peak, on May 22. A blast of hot gases flashed down the slope, which was
already partially devastated by the mudslide. The searing flash knocked down

every tree in its path and turned the forested hillside into a barren waste. The great cloud from the eruption deposited a thick layer of ashes as far away as Reno.

Most of the lakes, oddly enough, have neither outlet nor inlet, but are isolated depressions carved, for the most part, by ancient glaciers. They are kept filled by melting snows at a rate about equal to loss through evaporation and seepage. Cut off from the rest of the aquatic world, they were barren of fish until stocked by man. Today, many of these landlocked lakes can barely (if at all) support fish through the severe winter weather—although others, usually the larger ones or those on natural streams, offer sport ranging from fair to excellent, depending on conditions and the fisherman's skill.

Tulare is a phantom lake. As recently as a century ago, it was the largest lake west of the Rockies; now, farm tractors kick up great clouds of dust where paddlewheel steamers and sailing vessels once navigated. But, like phantoms, Tulare Lake and its smaller sisters—Buena Vista, Kern, and Goose lakes—sometimes rise to haunt those who have disturbed their beds. Fed by the Kern, the Kings, and other rivers that pour down from the southern Sierra Nevada, these lakes once covered much of the upper end of the San Joaquin Valley. Great quantities of alluvial soil that washed down from the mountains formed natural dams for the lakes that rose behind them.

In 1862, Tulare Lake was described as extending 60 miles north and south, as being 37 miles at its widest, and as covering 800 square miles. This shallow inland sea sheltered an abundance of aquatic life: trout that weighed up to 40 lb; perch, salmon, and sturgeon in quantities that later supported a commercial fishing industry; and terrapin that were served as delicacies in San Francisco restaurants. Ducks and geese were in such abundance that a single shotgun blast could bring down a brace of these waterfowl, and great herds of elk and deer grazed there.

Kern, Tulare, and Buena Vista lakes would wax and wane with seasonal variations in rainfall. Receding waters would expose alluvial soil which produced abundant crops, especially when irrigated from artesian wells or from any of the miles of sloughs that wound through the deltas of the Kern and Kings rivers.

A series of dry years and increasing irrigation gradually shrank the size of the lakes in the latter part of the last century. In 1880, the state legislature passed an act permitting settlers to buy reclaimable marshland for $2.50 per acre, $2 of which would be refunded if the settler spent a like amount on levees or other reclamation work. This touched off a land stampede.

Six times since records have been kept Tulare Lake has dried up entirely, and even in normal years it is a small and hard-to-find reservoir bordered by high levees. In the past, periodic floodings would return the lakes to life, causing considerable property and crop loss in the intensely farmed areas.

Dams on the Kings and Kern rivers have held much of these floodwaters in check in recent years, but the runoff from heavy snows in the 1970s inundated these old lakebeds, and Tulare Lake has spread over some 90,000 acres.

Spread across the Salton Sink in the torrid Colorado Desert is the blue expanse of California's largest lake, the Salton Sea. There is very little about its appearance that hints of days when the lake was being formed and when a large chunk of California was teetering on the brink of ecologic disaster.

Eons ago, the Gulf of California extended into the Salton Basin. But the silt-laden Colorado River built a sand dike across the Gulf, making the upper end a dead sea. After the salt water evaporated, the below-sea-level basin was filled from time to time by the capricious Colorado River; the river would then turn away again and the ancient lakes would dry up.

Early explorers found the area a scorching desert, swept by sandstorms. Emigrants to California, in 1849, wrote that "the only signs of human habitation were the bones of animals and men scattered along the trails."

Promoters renamed the Salton Sink "Imperial Valley," and investors and settlers were sought from all over the nation. More than 10,000 settlers had arrived by 1904. The sweltering basin, when provided with water, proved to be a vast natural hothouse. Grapes, melons, cotton, alfalfa, and other crops fairly sprang from the soil.

But trouble came early to this Eden. In 1904, heavy silting of the intake canal caused a water shortage. To meet this problem, the company decided to make a new cut in the river levee in Mexico. It was a fateful decision.

An unprecedented winter flood season prompted the California Development Company to throw a dam across the cut in March, 1905. It was washed away by spring floods, as was a second dam. The cut became a mighty gap through which roared 90,000 ft^3 of water per second. The ancient lake was revived, and it was feared that the whole valley might be doomed.

Left behind was the bankrupt shell of the California Development Company and the Salton Sea. Fed by the runoff of irrigation water, the lake is now saltier than the ocean and growing more so each year. Mud pots and fumaroles beside the lake indicate the presence of geothermal steam, which could possibly be used to generate electricity.

Cupped in a great, glacier-carved bowl of granite, Lake Spaulding stands at the headwaters of California history. Yuba and Emigrant gaps, through which the "49ers" struggled on their way down from Donner Summit, overlook the lake. The waters which feed it were once used for working some of the richest gold fields in the state. The network of canals and flumes linking Spaulding's watershed with the mines were engineering achievements that rivaled the Southern Pacific Railroad and Interstate 80, both of which now cross the mountains close by.

Water was needed to wash away the sand and gravel in which the gold lay buried. As stream-side claims were exhausted, canal companies were formed to bring water to "dry diggings," and many of the reservoirs and canals which they developed became the seeds of Pacific Gas & Electric Company's hydroelectric system. Most extensive and spectacular of these properties was the Rock Creek, Deer Creek, and South Yuba Canal Company, formed in 1854.

The South Yuba Canal Company (as it was later known) developed a series of reservoirs, among them Meadow Lake. The latter was the site of Summit City, where a boom town of 5000 persons mushroomed in 1865–66. When the gold strike petered out, the new city—complete with 13 hotels, a stock exchange, palatial saloons, and fine homes—was abandoned almost overnight.

Of Mono Lake, Mark Twain said "This solemn, silent, sailless sea—this lonely tenant of the loneliest spot on earth." Sometimes called the "Dead Sea of America," Mono Lake is still an impressive sight, although its shores are no longer lonely and its waters have receded considerably. The great escarpment and snow-capped peaks of the Sierra rise abruptly from beside its western shore. In other directions lies the haunting emptiness of the desert landscape of eastern California and Nevada. Although the lake covers about 100 square miles, no fish live in its bitter waters and boats rarely disturb its surface.

Mono Lake has no outlet, and much of the water from the mountain streams that feed it has been diverted to the Los Angeles aqueduct. Evaporation has concentrated its chemicals and shrunk its shoreline. Along its edge rise strange tufa towers, knobby white formations standing like so many Lot's wives—coral-like lime deposits of tiny calcareous algae.

Mono Lake has two treeless islands that are as unusual as the lake itself. One is small and black, the other is large and white. The former, Negit, is a volcanic cone of cinder and lava; though separated from the ocean by a mighty mountain range and 100 miles, it is a nesting place for thousands of seagulls who fly in from the coast every April and return every October. The other island, Paoha, is 2 miles long, and emits occasional wisps of steam from hot springs and fumaroles.

Clear Lake, the largest natural lake lying entirely within California, has been a resort area for thousands of years. Its hot springs and mineral waters, its fish and mild climate, its islands, and its mountain have combined to lure people to its shores since long before the first white settlers stumbled into this mountain-rimmed basin in the Coast Range.

Clear Lake is 18 miles long and 8 miles across at its widest point, and is a geological curiosity. It was once two lakes, the upper one draining into the Russian River through Cold Creek, and the lower one emptying into the Sacramento River via Cache Creek. When a lava flow dammed Cache Creek, the lower lake overflowed into the upper. The upper outlet was then choked off by a mammoth landslide. The water level rose even higher and eventually broke through the lava dike on Cache Creek, carving the present outlet.

For centuries, tribes of the Pomo Indians lived around the lake. They fished in canoes made from willow poles interwoven with tules, and the women made baskets fine enough to hold water. There was plenty of obsidian for spears and tools. They bathed in the hot springs, relaxed in long sweathouses, and gambled and played games on the small islands that dot the lake. Volcanic Mt. Konocti, with its eerie caves and its fissures with strange drafts, was considered sacred ground.

Deposits of sulfur and quicksilver around the lake were mined through the years. Borax extracted from Little Borax Lake met the entire needs of the nation from 1868 to 1873, when the discovery of the great borax deposits in Death Valley ended this operation.

But the abundance of Clear Lake had its limits. Clearing and farming gradually released silt and nutrients into the lake, and eventually caused the once-blue waters to cloud with dirt and algae. By the 1940s, the infamous Clear Lake gnats had become a great problem, and three sprayings with DDT were tried over the next 8 years. The gnats developed an immunity to the lethal chemical, but fish and fish-eating birds and animals began to die off.

Today, the ecological balance of the lake has largely been restored by use of a short-lived pesticide to control the gnats and by the introduction of millions of tiny freshwater smelt. The latter feed on gnats' eggs and also use the nutrients which had caused the algae to grow. In turn, the smelt are food for birds and larger fish, both of which have made a major comeback. A researcher from the University of California at Davis commented: "It is the rebirth of a lake."

"Gold! Gold from the American River!" Today one can almost hear the echo of these words in the breeze as it blows across Folsom Lake. This large reservoir on the American River is in the heart of gold country. Just upriver is Coloma, the original discovery site. Downstream is the city of Folsom, with its gas-lit mall straight out of the Old West.

And buried beneath the blue waters of the lake is the site of one of the state's most ancient gold mining communities, Mormon Island. It was there that Sam Brannan opened his store, early in 1848, for the miners whom his news had attracted. Sam soon became the state's first millionaire, but his luck turned and he died penniless in 1889.

Folsom Lake was created by the construction of the Folsom Dam project by the U.S. Army Corps of Engineers. Completed in 1956, the project is actually several dams: the big concrete main structure built across the American River, and two wings dams, an auxiliary dam, and eight dikes that fill in the saddles in the rim of the reservoir. The reservoir is enfolded in the foothills at the edge of the Sacramento Valley plain.

Lake Berryessa is nestled in a low and hidden valley between two spurs of the Coast Range in Napa County. It is a picturesque and relatively new man-made lake with some unusual properties. With water temperatures ranging from around 70°F at the surface to 40°F at the bottom, the lake houses a variety of both warm- and cold-water fish. Trout, black bass, catfish, kokanee salmon, steelhead, blue gill, and crappie are available.

Created by a dam built by the Bureau of Reclamation, it has 168 miles of shoreline, more than Lake Tahoe, and is one of the largest man-made lakes in California. Berryessa was once a valley so rich and idyllic that the nomadic Pomo Indians could live a life of ease off the game and natural vegetation there. In later days, farmers fought gun battles to protect their holdings, and claim-jumpers were a constant threat.

In the good old days life was sweet beside Honey Lake—if you lived to enjoy it. The lake, although alkaline and on the edge of the great Nevada desert, teemed with waterfowl. The hills sheltered an abundance of wildlife: deer, elk, bear, and rabbits, as well as hostile Indians and renegade white men.

Crops grew easily in the fertile soil, gold was there for the finding, and tax collectors were not welcome. The Honey Lake settlers packed six-guns and rifles, did for themselves what needed to be done, and earned (without effort, of course) the nickname "Never-Sweats."

Honey Lake is on the eastern side of the Cascade Mountains and Sierra Nevada. Originally included in Plumas County, today it is part of Lassen County. Old Peter Lassen was probably the first white man to see the lake, in 1850. He gave it its name because of the sweet, sticky substance secreted by aphids on the shore plants.

At times "lake" is as much a misnomer as "honey," for in some dry years the water evaporates, leaving only a 15-mile-long playa. But with or without water, Honey Lake was an early attraction to settlers, and around it revolved a shooting war between California and Nevada.

The largest man-made lake in California is Shasta, a great four-fingered reservoir encompassing vast chunks of the state's geography and history. From the concrete dam that forms it to the rivers that feed it, everything about Shasta Lake is on a grand scale. It has 365 miles of shoreline, more than San Francisco Bay. It can hold enough water to flood an area the size of Connecticut to a depth of 1½ ft. Only Hoover Dam is higher than Shasta Dam, and the only concrete dam in the nation that is more massive is Grand Coulee Dam.

Shasta Lake was formed in 1943, when the floodgates were shut, even while the upper works of the dam were still being completed. Keystone of the

Central Valley Project, the reservoir is on the Sacramento River, whose roaring upper reaches were called "Destruction River" by early explorers.

Also feeding the lake, and forming its other three fingers, are the McCloud and Pit rivers and Squaw Creek. Situated at the north end of the Sacramento Valley, Shasta Lake lies at the junction of the Coast and Cascade ranges. Towering nearby is majestic Mt. Shasta, whose melting snows and glaciers feed the many streams and rivers of the lake's vast watershed.

The lake is set in a land of ancient lava flows and active volcanoes, of ice caves and spectacular limestone caverns, some of which, e.g., Lake Shasta Caverns, rival in wonder any of the more famous caves and caverns in the nation.

In the rugged confines of the great Feather River Canyon, human dreams and achievements have always tended to match the grandeur of the scenery. Lake Oroville is no exception. It rises behind the highest dam in the nation and is the keystone of the multi-billion-dollar State Water Project. Lake Oroville has 167 miles of shoreline and extends as much as 15 miles up the three forks of the Feather River, the largest to flow from the Sierra Nevada. Its waters are carried through aqueducts and tunnels to water-shy areas of the state—they will eventually reach almost to the Mexican border.

The story of Lake Oroville could begin in 1817, when Luis Arguello led a Spanish exploration party into the canyon and gave the river its name (Las Plumas in Spanish). But it was in 1848 that the Feather River region leaped into prominence. John Bidwell had panned enough gold from the river to whet his interest, and on July 4, 1848, he and his companions made a rich strike at what became known as Bidwell Bar.

By the fall of 1849, a shanty town called Ophir had sprung up beside the Feather River. Its name was changed to Oroville in 1856, a year when more promising gold strikes elsewhere had largely depleted the town's population. The Feather River yielded riches only for a relatively short period.

Frank McLaughlin arrived in Oroville in 1879, with a commission from Thomas A. Edison to look for platinum to be used in filaments for his electric lamps. McLaughlin found no platinum, but, in 1880, he came up with a plan to bore a 2-mile tunnel through a bend in the river, divert the water through it, and so lay bare 11 miles of stream bed.

But when the river was diverted, the results were disappointing. A few years later he formed two other corporations to lay bare another stretch of the river just above Oroville, this time by building a 2-mile-long rock-walled diversion channel. It took about 4 years to complete. When the river diverted, the results were equally dismal.

Engineers had long recognized the power potential of the Feather River and, in 1901, preliminary steps were taken to capitalize on it. The tunnel was used to carry water to a big power plant. The undertaking was made more difficult by the San Francisco earthquake of 1906, which scared off investors from the East. The surveying and building of the Western Pacific Railroad through the Feather River Canyon plotted the best grade for this low-level route through the Sierra. In the fall of 1905, work began on the road from Salt Lake City to the Pacific by way of the Feather River.

Today, the waters of Lake Oroville cover both Big Bend Powerhouse and much of the original railroad and highway routes. More than 20 miles of each had to be relocated in the Oroville Dam project. The dam itself contains 80 million cubic yards of gold-dredge tailings and clay. Two powerhouses below the dam develop a capacity of 725,000 kW.

The big dam cost $414 million, 16 times the estimated value of the gold taken out of the Feather River. Completed in 1967, it has eased the danger from the disastrous floods which periodically swept the Sacramento Valley. With 15,400 acres of water surface, the lake is a mecca for recreation seekers.

Although the large-scale hydraulic mining of gold lasted less than two decades, it left an impact on California that is still being felt today. Before the practice was outlawed in 1884, miners had washed away whole hillsides with powerful jets of water. The mud and debris that flowed downstream raised riverbeds, caused towns and farms to be flooded, disrupted riverboat traffic, and laid waste verdant mountainsides and fertile farmlands.

At the same time, hydraulic mining provided the stimulus for the construction of the great water-gathering systems in the Sierra. Low-cost hydroelectric power owes much to the waterwheels and nozzles invented by the gold miners. And the world's first long-distance telephone system was built to serve these hydraulic mines.

Today, the largest of the old hydraulic mines is a state historic park, Malakoff Diggins. Situated on San Juan Ridge between the South and Middle forks of the Yuba River, the park includes the ghost town of North Bloomfield, once populated by 1800 persons and boasting seven saloons and two breweries. Nearby are the diggings: a 1600-acre, water-filled pit surrounded by painted cliffs that have been weather-sculptured into a surrealistic landscape.

Described as "one of the most perverse, unruly and unpredictable bodies of water in California," Lake Elsinore today glints placidly in the sunshine. Its calm waters give no hint of more tempestuous days, when the lake was given to spewing up dead fish, turning blood red, emitting evil smells, and even disappearing entirely.

Cupped in a pleasant valley on the eastern side of the Santa Ana Mountains, Lake Elsinore is fed by the San Jacinto River, which rises in the nearly San Jacinto Mountains. Hot sulfur springs bubble beside the lake.

Lake Elsinore, settled by the Luiseno Indians, may have been sighted as early as 1769, by scouts of Portola's expedition pushing northward in search of Monterey. More likely it was Don Pedro Fages, exploring inland in 1782, who was the first to see the lake, once southern California's largest. A branch of the southern Emigrant Trail later cut through Elsinore Valley.

Franklin H. Heald first saw Lake Elsinore when he climbed Mt. Baldy in 1880. Intrigued by this little-known body of water, he and two partners bought almost 13,000 acres of land around the lake in 1883, a dry year, when the normally pleasant lake was little more than a swamp. Heald staked out a townsite and sold lots. A deluge, in 1884, brought the lake back and Elsinore flourished. The hot springs and mineral waters brought visitors from all over, and Southern California's land boom was on. The arrival of the railroad added to the prosperity.

But the fortunes of Elsinore waxed and waned with the level of the lake. After long, dry cycles it would shrink. Then the lack of oxygen brought periodic fish kills, and sulfur vapors would flavor the air. The lake once unaccountably turned red after an earthquake. Despite the dry years Elsinore remained an attraction.

Just when things were going well, there would be a series of dry years and the lake would dry up. In 1949, a landowner who had purchased the lake bed began fencing it in. A park district was formed, the lake and its bed were purchased with state aid, and steps were taken to insure a dependable supply of water to keep the lake filled. On February 1, 1964, water from Lakeview Reservoir was "turned on," and Elsinore—which had once again gone dry—began to fill. A state park was dedicated beside the 3-mile-long lake.

From above, Donner Lake and the floor of the Truckee Basin—surrounded as they are by the eastern ramparts of the Sierra Nevada—look like a giant amphitheater. The stage was set eons earlier, when primeval forces tilted up the mountain range and subsequent ice sheets covered all but the highest peaks. When the final glacier retreated, it left on the floor of the ice-carved valley a natural dam, behind which the waters of Donner Lake rose. During the years 1845 and 1846 a flood of emigrants used the route that passed beside what was called Truckee Lake.

High in the wilderness area of the San Rafael Mountains, where a branch of the Santa Ynez River rises, is one of the last refuges of the great California condor. The picturesque Santa Ynez Valley is a land of contrasts. The old-world Danish village of Solvang sits not far from a freeway. In the heart of the valley is Lake Cachuma, a great blue reservoir that lies where many once thought a lake had no business to be.

Cachuma, together with two smaller, earlier reservoirs, furnishes water for Santa Barbara and the surrounding coastal plain. When the project was proposed in the 1930s, it ran into a storm of opposition from those who declared that the water would be too costly, that the often-meager flow of the Santa Ynez River would never fill a major reservoir, and that it would interfere with downstream rights. In 1949, voters approved the project, to be built by the Bureau of Reclamation under agreements with local agencies.

Work started in 1950, with the blasting of the first rock for a 6-mile tunnel that would carry the lake's water through the Santa Ynez range to the coast. The drillers moved rapidly through chalk and sandstone, working on both sides of the mountain. Then they ran into trouble: they met torrents of hot water, hydrogen and sulfide gases that would gag the strongest miner, explosive methane gas, flowing sand that oozed in faster than it could be removed, and unstable rock that buckled steel supports.

The Tecolate Tunnel soon became known as "the toughest tunnel in the world." Miners rode to work in mine cars filled with tepid water to keep their body temperatures from rising to dangerous levels. Up to 13 million gal of water a day, much of it 112°F or hotter, steamed into the tunnel, rapidly rusting all iron and steel, constantly shorting out electrical connections, and making it difficult to breathe and almost impossible to work in the 100% humidity.

The tunnel proved to be a horizontal well, and, in 1951, water from it was piped to Santa Barbara and Montecito to help relieve shortages. It required 6 years to finish, at a cost three times the original estimate.

The dam forming Lake Cachuma was finished in 1953, and the lake filled for the first time in spring of 1958. The completion of Tecolate Tunnel and the arrival of Lake Cachuma water was a great event for the parched coastal area. The lake, 8 miles long and about 1 mile wide at its widest, is a major recreation area for the region.

Gold Lake is the largest of 50 small lakes scattered over the glacier-scoured ridge between the North Fork of the Yuba and the Middle Fork of the Feather rivers. At 6400 ft elevation, the 27-mile-long lake is a central feature of the Lakes Basin area.

In 1850, it was a magnet for every loose prospector in Northern California. Miners gave up rich claims to scramble through the Sierra in search of this lake which, excited Argonauts claimed, had a shoreline and bottom of almost solid gold. It was Thomas R. Stoddard who stumbled into civilization carrying a sackful of nuggets and telling of a lake lined with gold.

Miners abandoned $100-a-day claims to follow Stoddard back to the lake. Farmers dropped their plows and storekeepers shut up shop; Marysville was almost depopulated. For weeks, thousands of searchers wandered the high country above the Yuba and Feather rivers. Gold Lake was given its name sarcastically when Stoddard stopped there briefly, saying it looked somewhat like the golden lake he was vainly trying to find. Gold Lake was a glorious humbug, though it is said that Gold Lake's bonanza is still there, but covered by great rock slides.

Meadow Lake, on the opposite side of the Yuba River canyon, was another humbug, but it was a caper pulled by nature and not by man. Gold was indeed there, but retrieving it was another problem. The lake is one of the many natural and man-made lakes lying just below the summit of the Sierra Nevada, near Donner Pass. Nature's jest became apparent when it was found that the gold was so firmly locked in veins or iron ore that no known process could separate them. Only at the surface, where the exposed ore was oxidized and had decomposed, could the gold be recovered.

Meadow Lake is part of the historic Yuba–Bear hydroelectric system. Gold Lake was acquired by Pacific Gas and Electric in 1927, in connection with a proposed hydroelectric development on the Middle Fork of the Feather River, a project that never materialized.

The ancient Merced River twists its way out of the rocky embrace of Yosemite Valley only to be captured again farther downstream by two recent but perhaps lesser-known reservoirs: Lakes McClure and McSwain. Enfolded in the foothills of the southern Sierra and surrounded by relics of California's colorful past, these two lakes store water for the Merced Irrigation District. They provide flood control, irrigation water, power, and recreation.

Most of the waters that feed these lakes rise in the granite fastness above Yosemite Valley. They tumble into the canyon in cataracts to form the Merced River and flow into the San Joaquin River, only 60 miles due west. On its way, the Merced exposed the southern end of the rich vein of gold that streaked the Sierra Nevada: the Mother Lode.

The Merced River was a pathway to the gold fields, and also to Yosemite Valley. Work on the Yosemite Valley Railroad began in 1905, and from 1907 to 1945 the line carried freight and passengers between the mountains and Merced. Here were some of the earliest developments of water power. In 1854, a water-powered grist mill and later a woolen mill were built at Merced Falls.

The significance of the area today is the multipurpose water development of the Merced Irrigation District. Formed in 1919, for the purpose of acquiring several existing canal properties, the MID started construction of Exchequer Dam in 1926, on the site of an old timber-crib mining dam.

In 1964, a new and larger Exchequer Dam was built, enlarging the capacity of Lake McClure to more than 1 million acre-feet. A smaller dam was built downstream to form Lake McSwain. Two hydroelectric plants provide a capacity of 89,100 kW. These reservoirs provide irrigation water and flood control for some of the richest farmland in the San Joaquin Valley.

8–4 THE EVERGLADES

To Miami, it seemed to be the only logical, sensible course of action. Thirty-nine square miles were purchased in central southern Florida, $13 million were spent, and a landing strip for training flights was constructed. Florida needed a new jetport: the site beside Everglades National Park was reasonable—it would be within swift reach of the booming cities on both coasts once expressways had been constructed.

No federal funds were granted for the proposed jetport in the Everglades. The port, with its satellite industries and residential developments, would have to be built elsewhere. And all to save an apparent wasteland—a super-swamp, an endless sea of shallow-water sawgrass—from the pollution of jet sound and jet contrails, and from the attendant on-ground sewage and industrial waste; all to save a 1.4-million-acre mega-bog, abounding in the horrors of the Everglades: alligators, poisonous snakes, clouds of mosquitoes, and huge, biting flies.

Natural assets and wildlife preserves have been rescued before. What was new here was the magnitude of the work already done, the money spent, the solid expectations suddenly rejected. The fiat had been made against tremendous commercial investment and popular demand and need, against its hotels and supermarkets and other cultural artifacts that would attract the whole world of air travelers and become 50–500 times as great a source of profits and taxes as the million or so tourists who now visit the Everglades each year.

One logical argument that the jetport opponents had been able to summon up was easily expressed. The aquifer from which the urban sprawl of coastal cities draws water might have been polluted by the development. These many trillion-gallon stores of ground water lie only 100 ft below the porous rock of southern Florida. Already that natural storage cistern had been diminished by saltwater incursions to the east and west, caused by activities such as canal digging and drilling.

The Everglades are, ecologically, unique on the planet and extremely complex. A map of Florida will show why. The southern third of the peninsula will be marked "Everglades." This vast wetland is, in fact, three kinds of swamp. The northernmost begins at Lake Okeechobee, a shallow body of tepid fresh water more than 700 square miles in area. The lake is (or was) the "head" of the Everglades supply of slow-flowing water, called a "river of grass." The first segment of the Everglades is the Big Cypress Swamp, although all of the big trees have been cut and most of the cypress had always been stunted and small, though often very old. Next comes the sawgrass region of brownish "grass" standing in shoal water and broken only by jungle domes, called hummocks. The sawgrass is an abrasive sedge, and a man trying to bull through it would soon be stripped of clothing and of skin.

The third swamp is a mangrove forest, where labyrinthine waterways twist and branch and open into secret lakes. Mangrove is impassable for any distance, as its roots and stiff, intertwined branches stand in slow-moving water that becomes brackish, then salty, and, finally, seawater. All three swamps compose the Everglades, which occupy the whole peninsula from edge to edge, a swamp of more than 5000 square miles soaked by a river that is the world's slowest, shallowest, and, perhaps, least dependable.

It is possible to build in the Everglades. Many developments already encroach on them. The vast wasteland could be turned into a megalopolis with modern machinery. The excavation of fill for building sites would merely leave large stretches of water canals and lakes, an effect that could rival Venice and be huge enough for 10 million or more inhabitants. If the Everglades were to vanish beneath an aquatic super-city, humanity could exist without missing a thing, save for a few lowly creatures.

The Army Corps of Engineers has constructed "water conservation" areas to the north and has diked ponds to store the rainy season's deluge. They have failed to supply the park with sufficient water in dry periods to sustain its flora and fauna.

The mucklands south of Okeechobee sustain sugarcane and winter-vegetable enterprises. The first farmers to see the black, friable soil thought it as fertile as the dark earth of the Midwestern prairies. It is composed of humus, leached of minerals, that requires fertilization, and, as it is kept dry by ditching and canals, it burns up and blows away, so that drainage systems must be constantly deepened until bedrock is reached. As the digging deepens, saltwater intrudes from the surrounding seas.

You could lead an informed citizen into the Audubon Society's Corkscrew Swamp and show him native hibiscus in bloom and stands of cypress bigger around than a dining-room table, and you could spend as much time as he would allow pointing out the rare, the gorgeous, the irreplaceable, and the possibilities for all humans in these resources. But his reaction might be: Trees are lumber, and a quagmire is a stinking breeder of pests. Prairies are to plough, rivers are for sewage and waste disposal, lakes are for dumping and transport and boating fun, scenery that lies over ore or petroleum deposits should be removed or drowned in muck, and all swamps, of course, are for draining. This is land improvement and, unfortunately, even those who oppose it usually do so for superficial reasons—to watch birds, or hunt them.

Our great illusions continue. Nature cannot be conquered or controlled, as we believe, because we are not fully in charge of its forces, and never will be. Who commands the wind and rain, the green plants and photosynthesis, the birds and insects, and the seven seas? Nature rules, exclusively and forever.

The Everglades offer a perfect illustration of what the human race has not yet begun to face: No one owns anything, and no one enjoys more than the temporary use of his presumed possessions. We do not own the Everglades or any part of that strange land, even though we may possess a deed. We are allowed its use, but all that we ultimately own is what our individual skins contain. To save ourselves, we must preserve whatever chains of life are essential to our own.

The value of the federal decision against a jetport rests in the symbolism of the act. If the symbol is understood, its worth will be immeasurable. For we will soon be compelled to forego countless developmental opportunities and to change our plans for many others, not for the specific, if obscure, goal of swamp salvage, but for the greater, though scarcely comprehensible reason that we know too little to risk so much.

8-5   AVERY ISLAND, LOUISIANA

New Iberia is bayou country, a region of giant gnarled oaks, Spanish moss, and sleepy alligators. You pick up Louisiana Highway 14, drive a few miles past rice and sugarcane fields simmering in the warm sun, make a left onto Louisiana Highway 329, cross a small bridge, and come to a stop at a tollgate at Avery Island.

Avery Island is a giant dome of rock salt, rising 152 ft above sea level, the highest point in southern Louisiana, surrounded by marshlands and the Bayou Petite Anse on three sides, and by a cypress swamp on the fourth. It presents acres upon acres of exotic tropical and native blossoms, trees, and plants; it is the site of the nation's first salt mine, and one of the largest; it is a bird sanctuary; it is the birthplace and manufacturing home of Tabasco pepper sauce; it is the center of a major operating oil field. Wildlife, mining, manufacturing, and petroleum operations exist in harmony within an area 2½ miles long, 1¾ miles wide, and 6 miles in circumference. Few of the thousands of visitors who come yearly to gaze on the wonders of Avery's Jungle Gardens and Bird City are aware that three major industries are flourishing just an egret's flight away.

Seventy-eight oil wells are currently active. Together, they yield 6900 barrels of oil and 15 million cubic feet of gas a day. The wells are mostly scattered among the bayous and channels that surround the island. Each well is capped with a low "Christmas tree" of valves and is encircled by a protective guardrail.

Millions of tons of salt have been taken from the Avery Island mine and countless millions remain. Beneath the thin veneer of soil lies an underground pillar of salt 2 miles in diameter that extends downward at least 20,000 ft. Excavations below the surface have carved out vast cathedral-like "rooms" up to 100 ft high, with ceilings, floors, and sidewalls all of rock salt. The mine workings extend ¾ mile across at the widest point. In the subterranean city, huge electric mine equipment looks like children's toys. Supervisors use jeeps to check operations.

On the eastern side of the island are cultivated fields of pepper plants. These provide the prime ingredient of Tabasco-brand pepper sauce, made only on Avery Island. Once the salt and pepper industries had been firmly established, the saga of Avery Island entered its next phase—the planning and development of Bird City and the Jungle Gardens.

The population of Bird City has grown to such an extent—to 100,000 or more birds during the summer nesting season—that it has been necessary to give nature a helping hand. Long, double-decked nesting platforms have been built on stilts out into the lake. Nearby, truckloads of twigs and brush are dumped each year for nest-making material. In this hospitable atmosphere, the long-legged, alabaster egrets have been joined by blue herons, black ibises, ducks, gallinules, and a wide variety of other water birds. There they thrive in a sanctuary that, remarkably, is a scant 200 yards from the Tabasco factory, 40 ft from a railroad spur that runs to the salt mine, and only inches from a paved road where trucks loaded with oil-well casing and equipment go clattering by.

None of this activity has disturbed the native wildlife. Alligators, mink, muskrat, raccoons, and opossums still abound in the marshes around the wells. Egrets and herons swoop down in the search for shrimp and minnows, and occasionally a deer wanders down from a nearby wood to see what is going on.

If ecologists could pool their knowledge with the rich data from all our sciences, we would still, alas, remain unable to say with full certitude what life forms and life systems are essential for human survival. We are simply novices in our understanding of the intricate, living infrastructure that supports our fragile species, and can no longer afford to risk losing any wild living form, lest one break in the planetary, life-sustaining system prove at last fatal to our civilization.

Nature has never guaranteed human survival for any particular period of time. But in the natural order of the universe, there cannot fail to be points where the loss or breakage of a system might set in motion an inexorable act of nature fatal to humankind.

CIVILIZATION IN THE TWENTIETH CENTURY

9-1 OUR WASTE IN AN URBAN ENVIRONMENT

In New Jersey, the most urbanized state in the United States, all of the problems in the confrontation between civilization and nature are cast into sharp focus. Here are the dirty air, the traffic jams, the filthy streams, the littered beaches, the oil spills, the decaying cities, and the pressures of housing developments on areas that have been farmlands for centuries. Here is urban America; here is the future of most of America. It is the affluent who darken the skies, who foul the streams, who dump their garbage and waste off boats into Barnegat Bay. It is the affluent who complain that animal wastes on nearby farms offend their guests.

Waste comes in many forms—garbage, old cars and refrigerators, plastics, throw-away cans and bottles—but in New Jersey, the type that is assuming increasing importance is animal waste. New Jersey is a paradox: it has the highest percentage of people per square mile in the United States, and yet it is also largely an agricultural state, with nearly two-thirds of its land area still in farms and forests. The reality of farm life often comes as a shock to the former city dweller. The pastoral scene admired in paintings or photographs hardly prepares an urbanite for the odors that are likely to come wafting over a barbecue party of a warm spring evening.

As the passing years have seen the cities spread into rural areas and open land shrink, the problem of efficient agricultural waste disposal has become an important factor in developing the strong and viable agriculture which is basic to the continuing progress of our economy. New technology and new development are the keys to maintaining a balance between rural and urban America as they find themselves living in ever closer proximity to each other.

On a tonnage basis, agricultural wastes represent approximately two-thirds of the total of 3.5 billion tons of solid waste produced annually in the United States. One-and-a-half billion tons are animal and poultry waste, and 550 million tons are leftovers from the marketable portions of farm crops. It is estimated that more than 5000 tons of such waste are produced in New Jersey each day.

It is possible to visualize a time when all of the farmers will have left New Jersey. If this should happen, milk, eggs, fruits, and vegetables would all need to be imported, with attendant price increases. No longer would the "Sunday driver" be able to stop at a roadside market and load his car with tomatoes, green peppers, peaches, blueberries, and sweet corn.

However, cooperation and understanding are a two-way street. The farmer must also become aware of the homeowner's problem. It is not sufficient to say "I was here first." These new neighbors of the farmer are the people who buy his products.

Since 1933, the Walker–Gordon farm in Plainsboro, New Jersey, which now has a concentrated herd of 2300 cows and calves, has solved the manure disposal problem through the dehydration, processing, and sale of an organic soil conditioner.

Until recently, that is. Changes in animal housing and manure handling to effect greater efficiencies in the operation, as well as urban housing encroachment, have created problems. Close neighbors—near gas odors from the processing stack, aerobic storage-pit odors, and so on—have made complaints which threaten the existence of a solution to the solid waste handling problem and farm operation.

In the middle of 1973, the Environmental Protection Agency's publication, the *EPA Log*, stated that burping cows must rank as the primary source of air pollution in the United States. EPA estimated that American cows burp 50 million tons of hydrocarbons into the atmosphere annually. A further statement was that ten cows burp enough gas in a year to provide for all the space and water heating, as well as cooking, for a small house. The Wyoming Department of Agriculture, when informed of this study on cow pollution, wondered whether the work would be extended to all animals, domesticated and wild, as well as to man.

Animal waste is processed into a relatively stable material suitable for soil conditioner or landfill. Composting greatly reduces the volume of the waste to as little as one-tenth of the initial volume. Machines in Cape May compost a mixture of garbage and hog manure. A major share of Philadelphia's garbage is sold to New Jersey farmers for feeding hogs. If Philadelphia lost this source of revenue, an initial expenditure of $7,500,000 would be required for compacting equipment alone, plus yearly labor and maintenance costs.

And although the garbage problem can to some extent be dealt with as Philadelphia has done, the animal waste problem remains. It is hoped that the composting machine will help resolve the problem, thereby allowing additional swine operators to come into New Jersey along with the attendant new feed plants and jobs. Hog farming currently adds $15,000,000 a year to the state's economy.

The oldest waste disposal method known to man is simply to spread the waste product on the ground, and the newest is the subsod injection method. Others include plow–furrow–cover, irrigation, lagooning, digestion, sanitary landfill, and incineration. There is no single solution: all have their place.

An example of what can happen when organic material is not returned to the soil can occasionally be seen on dry, windy days in New Jersey. At certain locations and under certain conditions, the area bears a strong resemblance to a Midwestern state stricken by a dust storm.

Complaints about the automobile as a polluter of air, a gobbler of space, and a despoiler of landscape have come to be familiar. They were familiar in turn-of-the-century America, as, for example, in the book *America Adopts the Automobile, 1895–1910*, by James J. Flink of the University of California. But in 1895, the first issue of the monthly, *Horseless Age*, welcomed "a giant industry struggling into being" and "the new civilization that is rolling in with the horseless carriage." "It would be," *Horseless Age* predicted, "a higher civilization than the one we now enjoy." This spacious and optimistic tone proved to be typical of many subsequent comments and predictions about the automobile.

One major reason for the early enthusiasm about the automobile was that it promised to rid towns and cities of clattering ironshod hooves and ubiquitous horse manure. Critics of the automobile sound as if they believe that its advent spoiled everything, rudely shattering the peace and fouling the pristine air of an Eden-like land. At the turn of the century, the horse constituted *the* urban problem.

"The nuisance will not be wholly abated," thundered *Horseless Age*, "until the great beasts whose refuse litters the streets and fouls the atmosphere in our populous centers are banished from the cities." *Scientific American*, in July, 1899, looking forward to the banishment, wrote:

> The improvement in city conditions by the general adoption of the motor car can hardly be overestimated. Streets clean, dustless and odorless, with light rubber-tired vehicles moving swiftly and noiselessly over their smooth expanse, would eliminate a greater part of the nervousness, distraction, and strain of modern metropolitan life.

Americans also foresaw a broad range of social benefits. There had been growing concern, for example, that a food shortage might result because so many people were flocking from farms to cities. The car, it was reassuringly said, would keep people on the farm, and horses as well.

Automobile enthusiasts overestimated the benefits to come (and also underestimated the social costs), partly because they possessed the optimism of a nation as yet unaware that its resources were limited. They multiplied the gains they could see for an individual by the number of eligible individuals. What sabotaged these forecasts was the failure to grasp the principle that as more and more people acquire cars—or boats, or second homes, or winter vacations, or incomes large enough to hire household help—the advantages of possession are diminished. The exercise of freedom by many limits the freedom of all.

Before we are led to believe that man will be buried under his own solid wastes or that he will die of polluted air, let us look at the record and draw our own conclusions.

Problems in solid waste operations have often been handled without employing all of the available information, and through the application of short-range remedies which have the advantage of low cost. Further, local interests have often prevailed over regional needs and planning. Most decisions have in fact been made chiefly on the basis of lowest cost, without considering effects on health, conservation, recreation, and aesthetics. To date, neither urban affluence, nor concentration, nor the automation and intensification of sales by industry, nor the great advances in agriculture resulting from mechanization, the use of pesticides, and biological control, have been tempered substantially by factors other than cost. It has often been remarked that the preservation of both public health and the aesthetics of the environment should be paramount in the management of industrial wastes, but too frequently this has been scarcely more than a dream.[1]

The cry is often heard that the earth's mineral resources are really quite limited. Metallurgists have known this for several generations, and mining practice has changed markedly from the times when ores ran 30% metal to those now in use when ores run less than 1% metal, ores from which a profit may still be successfully extracted. Similarly, when we now collect and smelt secondary copper-containing materials, we are also able to recover tin from the slags, zinc from the fumes, and byproducts in the form of selenium, tellurium, gold, silver, and platinum metals. Approximately one-half of our "virgin" copper arises from secondary materials. Lead from batteries is continuously recycled, and it has been estimated that before each new lead molecule has been dispersed so widely that it can no longer be successfully collected, it will have been recycled 300 times. Currently one-half of our annual aluminum supply is obtained from secondary aluminum, and this metal is handled by primary producers. Zinc wastes from steel-galvanizing processes are collected, transported, and converted back to zinc. Abandoned automobiles are dismantled, shredded, and classified into steel scrap useful to mills. The charges fed

[1]C. L. MANTELL, *Solid Wastes: Origin, Collection, Processing and Disposal*, John Wiley & Sons, New York, 1975, 1200 pp.

into electric steel furnaces are 50% or more scrap metal. Recycling in the metal industry, including nickel, tin, cobalt, cadmium, gallium, and the precious metals, has been part of the daily practice of the process metallurgist for generations.

Up until now it was almost axiomatic that "the system for managing solid waste must be economically as well as technically feasible." If we are now to satisfy our aesthetic desires, assuring the beautification of our landscapes, this dictum must clearly be revised to state that the method need only be technically feasible, whether or not it is economical. Some facilities or items of equipment may be technically suitable but expensive solutions to waste problems, occasionally to the point that communities are unwilling to pay a prohibitive price simply in order to take advantage of the available disposal technology.

Only the large cities now find incineration economically feasible, owing to the increase in the costs of original investment and operation everywhere mandated by legislation requiring improved air-quality standards.

Landfills, incinerators, compost plants, and biological oxidation units do not enhance the value of surrounding properties. There are even isolated examples where residential neighborhoods located near sanitary landfills are now planned for ultimate conversion to parks, golf links, and similar recreational uses.

Archeologists who have studied the ten cities of Troy in the Grecian Golden Era have concluded that the people of that time merely dropped their food scraps on the floor—bones and everything else—and went right on living on top of them. Gradually the floor level rose and eventually the door would not open. They adjusted the door and built buildings on the rubble until there were ten layers of cities. Pollution did not trouble the Greeks, nor the Romans, nor the Carthaginians. Nothing was changed very much during the Dark Ages, the Middle Ages, or the Renaissance. But with the industrial revolution, pollution and darkened cities became a sign of progress and were accepted as such. A century or so later we are no longer happy with the results of the industrial revolution and wish to change it, with a return of atmosphere, land areas, and resources to the way they were during the agricultural era.

Humans, of course, are no different from other living creatures except that they possess a brain and a skillful pair of hands. With these, we are able to use our invironment more effectively than other living things, and can more adequately protect ourselves from the dangers and the enemies in our environment.

During evolutionary history many other forms of life have emerged, grown, and multiplied. But some of them, like the dinosaur, ultimately vanished, either because they were unable to cope with their environment and its changes, or because they were unable to cope with new forms of life which became their enemies. Almost every form of animal must devour some other plant or animal in order to live. So far, man has been smart enough to protect himself from his enemies, which range from bacteria and viruses to man-eating lions and sharks, as well as the inanimate enemies of heat and cold, and fire and flood.

Farmers had to cut down forests, plow fields, plant, cultivate, and harvest crops, and process food for use. They had to feed, care for, and breed cattle, sheep, and hogs. In the old days, every member of a family had to work hard at these tasks from dawn until dusk.

They invented tools to assist in their tasks. Soon with their tools they became efficient enough so that not all people had to spend all of their time producing food. Some moved to the towns and built cities and carried out new tasks, earning enough money to purchase their food from the farmers. Thus modern civilization began.

Then began, also, the era of technology—the search for new materials in nature: stone, copper, iron, and other materials from which to fabricate better tools and better structures. New food products were discovered and developed. Machines were invented to do many tasks. Then, finally, came the age of energy, in which heat from the burning of fuel was converted into mechanical energy to relieve man, at last, from the back-breaking toil of former years.

The exploitation of nature and natural resources then began in earnest. A number of nature's materials could be converted into useful products, and new kinds of materials could be synthesized from those occurring naturally. Man kept increasing his knowledge through the systematic investigation of natural phenomena. He learned more and more about how to use energy, how to use materials, how to fabricate new things, and also, finally, how to cure many diseases and allay human suffering.

Some waste products were discharged into the air in the form of smoke and dust. Sewage and industrial wastes were washed away into the nearest stream or lake. Large piles of solid rubbish were dumped into the sea, burned, and buried in the land or scattered on its surface.

The human race has always been concerned with the environment. We have been exploiting it for only a short time, in geological terms, and in modest ways have been trying to preserve its best features, its beautiful landscapes, its clean air and water. Now, however, we must abandon the idea that all of these things can be done without cost or without great effort. Everyone must share the cost—consumers, taxpayers, and stockholders. To return the environment to its pristine state—whatever that might have been—is impossible; but we can improve it. To avoid all human impact on the environment is also impossible.

Many are deeply concerned about population and the environment. They see herbicides in trees, pollution in running brooks, radiation in rocks, and overpopulation everywhere.

In Calcutta, when the cyclone struck East Bengal in November, 1970, dispatches spoke of 15,000 dead, escalated to 2 million, and then dropped back to 500,000. The nameless ones who died left no trace of their existence. Pakistani parents replaced the population loss in just 40 days.

What killed these people? The cyclone or overpopulation? The Ganges delta is barely above sea level. Every year, several thousand people are killed there in ordinary storms. A delta belongs to the river and the sea; humans intrude at their peril.

The name, India, conjures up a scene of a village road down which a half dozen cattle are being driven. At the rear, carrying a basket, is an old woman, though she is not yet 40. The cows defecate in the road and the woman scoops up the warm feces with her bare hands, slapping it into her basket. Later, she plasters the wet dung on a convenient wall. The wall is not her property—she is too poor for that—but it is her territory, and will be respected by others. When the patties are dry she will use a few of them as fuel to cook her rice; the rest she must sell to buy more rice.

To Westerners, now rediscovering environmental housekeeping, the practice may be attractive. Certainly, the Indian roads are kept free of dung. In the long run the community would be better off if the dung were put back into the soil, but it is today's rice that needs cooking.

Overpopulation exhausts the soil and destroys the forests, thus producing a fuel shortage; then dung is burned and the soil impoverished, thus diminishing future crops and increasing poverty. Unrestrained reproduction can lock a culture into its poverty.

Every year, tuberculosis, leprosy, enteric diseases, and animal parasites are the "cause of death" for millions of people. Malnutrition is an antecedent of death; and malnutrition is connected with overpopulation.

People are reputed to be dying of respiratory diseases in industrial cities, because of the "need" for more industry. The "need" for more food justifies fertilization of the land. We refuse to blame deaths on overpopulation.

The environment is dirty. It has always been dirty—nature started to muck it up eons ago. Those who have been in the Blue Ridge know that the Blue Ridge is blue, and the Smokies are smokey, because of all the muck that comes out of all those pine needles. Our cities may be abominable in many ways, but those who know what the streets of a medieval or Renaissance town looked like—who know what London really looked like in the seventeenth century—also know that, comparatively speaking, New York is paradise.

The difficulties of the environment can be managed.

New York City is faced with enormous crises because of its burgeoning volume of solid waste. The streets are increasingly dirty, not in comparison with the large cities of other countries but by our own housekeeping standards (see Table 39).

Table 39. Control of Particulate Concentration in the United States

Year	Authority	Average particulate concentration (μg/m^3)
1931–32	Public Health Service (14 of the largest cities)	519
1957	Health, Education and Welfare Continuous air monitoring in 55 cities	120
1968	64 cities	96
1969	64 cities	92

Calcutta, India, is by far the dirtiest city in the world, with its unwashed millions of people, its beggars and caste untouchables; with a large share of its population living under sheds and bridges, a lack of sanitary facilities, manure-producing sacred animals roaming the streets; and with its stench, filth, flies, disease, and burning corpses at the river.

Flying into Rio de Janiero at sunset is beautiful, but to try and cross Avenida Atlantica, which is parallel to Copacabana Beach, at four o'clock in the afternoon is impossible, owing to the streams of cars, lined bumper to bumper. Rio has poverty-stricken ''flavellas,'' shacks of corrugated steel and refuse, without sanitary facilities, without services, and with, perhaps, only one light bulb per 20 inhabitants. These slums cover the hillsides and pollute the lakes and watercourses. New York does not have a water situation so critical that you end up, as in Rio, trying to take a shower on the fifteenth floor of a hotel only to discover that the water has been turned off, or, when you want to go down to the lobby, finding out there will be no power for the elevators for several hours.

In Lima, on the west coast of South America, the Andes might seem similar to the Rockies in the United States, but they are not. They consist of piles of pebbles. Associated with these in the industrial and suburban districts, particularly in the poor areas, is defecation in the streets by both sexes. Everyone of the seven parishes has a church, a mayor's palace, and a huge *supremarcardo* (supermarket)—usually an untidy one. Lima also has its slums.

In Bogota, Colombia, on the Avenida Jiminez, perfect gems of ancient cars are so numerous, and breakdowns are so frequent that, in comparison, the Long Island Expressway and the Los Angeles Freeways seem like lonely racetracks.

Mexico City suffers the problems of an unstable geology since it lies on the site of a former lake, and is subject to earthquakes and the resultant waste. Paseo de la Reforma is a showcase, but out in the surrounding areas one will see poverty, dirt, and filth.

In a similar way, Buenos Aires has the stretches of beautiful Avenida Alvear, but it does not take much travel to find its slums, poverty, and waste.

London is noted for its fogs, litter in its parks, and the low social status of the "dustman," who is the equivalent of the American garbage collector. The Midlands—Manchester and Liverpool—are areas where the need for jobs was apparently more important than the need for beauty and cleanliness, and they show it.

Tokyo is the most polluted city in the world because industry and manufacturing come first without regulation, labor is cheap, and the general viewpoint is "anything for money," with a resultant complete fouling of its own nest and its surrounding waters. Paris has the extremely dirty Seine River, with tourist areas and slums; Cologne has the strongly polluted Rhine River, with four lines of barge wastes, with chemical salt factories covering over half the city, and yet also has a cathedral and beautiful parks on the shore of the river.

In Rome, Ostia, and along the Italian coast, one cannot swim with impunity because of the muck. Italy is so polluted that it is being deserted by tourists. Venice is a sinking city of open sewers. The Italians never tear anything down. The Coliseum houses over 10,000 cats with their resultant smelly fecal waste.

And then there is the "greenhouse effect" theory, which claims that the buildup of CO_2 in the atmosphere will cause a temperature increase; throughout the planet, and that we will either drown in a tidal wave resulting from the melting of the polar ice caps, or will roast to death.

At the present level of CO_2 in the atmosphere—about 330 ppm—and at an annual rate of CO_2 increase of about 0.7 ppm, it would take about 957 years to triple the current level of CO_2. To bring the CO_2 concentration to a 1% level would therefore take about 12,850 years. (But such speculations have no more scientific validity than the prediction that a puppy dog, at his present growth rate, will be 15 ft long and weigh 990 lb at age 5.)

As far as the CO situation is concerned, EPA states that the combined total of man's activities, worldwide, accounts for around 2% of the CO production; all other living organisms, vegetable and animal, account for the remaining 98%.

Waste, from an engineering point of view, is matter or energy that, at the existing level of knowledge, cannot be re-converted into value. From an economic viewpoint, waste has zero cash value, at best.

Our astronauts have described earth, as viewed from the moon, as "the blue, water-laden planet." A very large share of the earth's surface is covered by seas. The human race is indeed puny in comparison with the Sargasso Sea, a waste burial place in the Atlantic. The bores, resulting in periodic reversals of strongly flowing rivers by tides in China, on the Amazon River in Brazil, the Moncton River in New Brunswick, and causing 50 ft changes in sea level at the Bay of Funday and 40-ft changes in the upper reaches of the Amazon and the Orinoco, 1000 miles from the delta at Belem, dwarf us. Here we retreat from destruction, or remain, at our peril, to see all our works become wastes.

Avalanches, associated with unstable snow masses, can be immensely destructive. So can cyclones, hurricanes, typhoons, and torrential rainstorms. One prophet of doom insisted that atomic explosions were changing the weather pattern of the earth. The National Fire Prevention Bureau, because of its losses in disasters, undertook to track down each one—locally, nationwide, hemispherically, and worldwide—and with the aid of computers, correlated these with atomic explosions. Statistical studies of the data indicated that atomic explosions did not create weather, but that typhoons, hurricanes, and cyclones caused the atomic explosions. Viva statistics!

After the engineer DeLesseps, with the financial aid of the Rothschilds, had built the Suez Canal, he wanted to build the Panama Canal. The French severely criticized the project, claiming that the waters of the Gulf Stream would escape to the Pacific, and its absence along their coast would make the winters in France frigid. DeLesseps failed in building the Panama Canal because yellow fever and malaria decimated his forces, enemies which Goethals and Gorgas eliminated before successfully building the canal. Was any thought ever given to "vanishing species" of disease-carrying mosquitoes?

Of course, nature itself over the eons has often altered ocean circulation and the climates of vast areas. A single hurricane or typhoon has the energy of more than a thousand of the largest atomic bombs. The energy released by such a bomb would fail to match the destructive force and waste-producing capacity of a cyclone.

Torrential rains perpetuate rain forests where trees, bushes, vines, and climbers are rampant, all seeking the sun. These "green hells" reach an unstable stress state and collapse, killing everything in their path. But the jungle returns in a short time, animals return, and another collapse takes place in a continuing cycle.

Theologians tend to avoid disputing the statement that engineers are not allowed to go to Hell at life's end. There is worry that the engineer, trained in thermodynamics, would find that the temperature differences between icebergs and frozen areas, and sulfur fires and burning areas, pictured by Milton in *Paradise Lost* and by Dante in his *Inferno*, could be converted into useful work, supplying energy to machines which would air-condition Hell or Purgatory, making them welcome places for real estate developers, and a theological impossibility.

Earthquakes, slides, and movements in faults bring only temporary attention to seismologists and geologists, but the scars of destruction and the waste of lives pass into history.

For centuries, the Greeks, Romans, Phoenicians, and Egyptians have recovered salt from seawater by solar evaporation. There are natural basins where the sun, wind, and sea carry on solar evaporation on a scale that makes human efforts seem truly puny. Visit the Rann of Cutch in India, a flat plain, 60 × 185 miles, where the southwest Monsoons deliver seawater which deposits several-foot thicknesses of salt. Think of the early isolated seas and the Strassfurt salt deposits 2500 ft thick, Great Salt Lake and Bonneville Flats in Utah, Searles and Lassen lakes in California, coastal bays in Peru, basalt and salt mountains in Colombia (made into an underground cathedral), salt mines (100,000 square miles) in New York and Michigan, to understand the magnitude of nature's wastes. We accept these as mineral areas and adapt—just as we make geyser areas and dinosaur burial grounds into national parks.

The Persian Empire, of pre-Mohammed times employed extensive systems of irrigation over vast areas that provided rich crops for its people, but natural changes converted these areas into deserts. Meteorological changes are decreasing the size of the Caspian Sea and reducing the sturgeon and caviar fisheries. The Imperial Valley in California, made into a desert by nature, has been irrigated into an agricultural marvel, but the Sahara, Gobi, and similar areas in Iraq, Iran, India, Chili, Pakistan, and the Far East show that natural forces are dominant, untroubled, and little disturbed by humans or their works. Were we able to transport our wastes, huge as they are, to one of these deserts, we could bury them with impunity for a thousand years and cease to worry about the problem. With such an approach we might, over a hundred years, be able to convert part of the desert into a modern Garden of Eden.

Again, if we were to transport our wastes to active volcanoes and have them consumed therein, we would solve all of our problems without adding to the pollution problems already created by the volcanoes. They possess a burning capacity far greater than all the incinerators we will build in the next century.

9–2 TECHNOLOGICAL GROWTH

Throughout history, the human race has overcome the natural obstacles of rivers by building bridges over them or tunnels under them. We have reacted to mountains in the same manner. Some of the major artificial features of the earth are listed in Tables 40, 41, and 42.

For those whose home is the United States, the most industrialized nation in the world, Table 43 allows reflection on the nature and extent of the management of natural forces by the engineer, the technologist, the scientist, the

Table 40. World's Major Engineering Feats
(*Courtesy National Geographic Society*)

Structure	Item	Location	Feature
Longest rail tunnel	Mont Blanc	Simplon, Switzerland	12.3 miles
Longest road tunnel		France–Italy	7.2 miles
Tallest dam		Nurek, USSR	Under construction
Biggest dam		Tarbela, Pakistan	186×10^6 cubic yards
Tallest building	Sears Tower	Chicago, Illinois	1,450 ft high; 110 stories
Largest building	Boeing Jet Assembly	Everett, Washington	Space: 205×10^6 cubic feet
Tallest structure	KTHI—TV tower	Blanchard, North Dakota	2,063 ft
Longest single bridge span	Verrazano—Narrows	New York, New York	4,260 ft
Highest bridge	Royal Gorge	Colorado	1,053 ft (above water)
Deepest mine	Gold mine	Kolar, Mysore, India	11,500 ft
Deepest well	Oil well	Beckham County, Calif.	30,050 ft
Great Pyramid of Cheops		Egypt	481 ft high, 755×755 ft at base; built of 2,300,000 blocks weighing 2.5 tons each
Great Wall of China		China	1,500 miles long, 25 ft avg. height, 15 ft in width at top, 25 ft at base

Table 41. Major World Tunnels

Name	Location	Length (miles)	Year opened
Mont Blanc	France–Italy	7.2	1965
Great St. Bernard	Italy–Switzerland	3.6	1964
Viella	Pobla de Segur Viella, Spain	3.1	1948
Mersey–Queensway	Liverpool, England	2.8	1934
Kanmon	Yamaguchi, Fukuoka, Japan	2.1	1958

Table 42. World's Major Railroad Tunnels

Name	Location	Length (miles)	Year opened
Simplon II	Switzerland–Italy	12.3	1922
Simplon I	Switzerland–Italy	12.3	1906
Apennine	Italy	11.5	1934
Gotthard	Andermall, Switzerland	9.3	1882
Lotschberg	Bern–Brig, Switzerland	9.1	1913
Mount Royal	Montreal, Canada	9.0	1916

biologist, and the physician. The United States enjoys the services of more of these, as judged by the membership of its technical societies, than all the rest of the world put together, and the consequences are a self-evidently high standard of living.

The eighteenth and nineteenth centuries, often referred to as the era of the industrial revolution, brought machines which relieved many of the burdens long associated with the provision of human habitation, food, clothing, and power. As a corollary, these centuries of change through technology ushered in the agricultural revolution of the twentieth century, which has been centered on genetic studies of plant breeding.

The twentieth century, within the life-span and memory of many people now living, has brought the development of conservation, national parks and their acceptance, the renaissance of attempts to reclaim arid areas on a gigantic scale, and flood control and the TVA, followed by similar national and multistate authorities.

In this century, also, we have witnessed the extremely rapid growth of electronics, radio, worldwide and automatic telephony and communication, and weather satellites. The air was conquered as a medium of transportation through the airplane. People may now circle the earth, if they wish, in either direction. We may reach the far ends of the earth over or under the oceans, through the atmosphere, or over the land.

Table 43. United States Geographic Facts

Item	Location	Area (square miles)	Height (feet)	Latitude, Longitude	Length (miles)
US, 50 states		3,615,211			
Land		3,548,974			
Water		66,237			
Largest state	Alaska	586,412			
Smallest state	Rhode Island	1,214			
Largest county	San Bernardino, California	20,154			
Largest city area	Jacksonville, Florida	827			
Northernmost town	Barrow, Alaska			71°18′N	
Southernmost city	Hilo, Hawaii			19°42′N	
Southernmost town	Nealehu, Hawaii			19°4′N 155°35′W	
Easternmost town	Lubec, Maine			66°59′W	
Highest point on Atlantic Coast	Cadillac Mountain, Mt. Desert Island, Maine		1,530		
Largest and oldest national park	Yellowstone (1872) and Wyoming, Montana, and Idaho	3,472			

Feature	Location	
Largest national monument	Glacier Bay, Alaska	4,381
Highest waterfall	Yosemite, California (3 sections)	2,425
	Upper Yosemite	1,430
	Cascades in middle	675
	Lower Yosemite	320
Longest river	Missouri–Mississippi	3,860
Highest mountain	Mt. McKinley, Alaska	20,320
Lowest point	Death Valley, California	−282
Deepest lake	Crater Lake, Oregon	1,932
Highest lake	Waiau, Hawaii	13,200
Coterminous 48 states		3,022,387
Land		2,971,494
Water		50,893
Largest state	Texas	267,339
Northernmost town	Penasse, Minnesota	49°22′N
Southernmost city	Key West, Florida	24°33′N
Southernmost mainland town	Florida City, Florida	25°27′N
Highest mountain	Mt. Whitney, California	14,494

Our ability to communicate was markedly augmented by Telstar I, launched on July 9, 1962, which, the next day, transmitted signals from Andover, Maine, and back, and to Holmdel, New Jersey, so that viewers in the United States, England, and France enjoyed the sight and sound of the same television program. Telstar I weighed 170 lb and circled the earth every 138 min.

Around the globe, people are reaching out by telephone to talk to friends, relatives, and associates with a convenience and frequency that could hardly have been imagined by those who witnessed the world's first commercial overseas call only 46 years ago.

In the intervening years, overseas communications have vastly improved. Telephone customers today can reach 98% of the world's 291 million telephones and can call beyond the United States, Mexico, and Canada to 244 nations or territories, including the People's Republic of China and places as distant as New Ireland or the Admiralty Islands.

In 1948, fewer than 800,000 telephone calls were made from the United States to nations abroad. In 1972, the total exceeded 39 million—a 50-fold increase, with volumes expected to rise to the 200 million level by 1980. Almost 200,000 calls will be handled between the United States and the Soviet Union, and considerably more between the United States and mainland China.

Color television programs transmitted via satellites enabled everyone with a receiver in the United States to see the Olympic games from Tokyo in 1964, from Mexico City in 1968, the winter games from Sapporo, Japan in 1972, and the summer games the same year from Munich, Germany.

Man in his home has seen, in color, at the same time it happened, the coronation of Pope Paul VI in 1963, the investiture of Prince Charles of England in 1969, and the visits of President Nixon to China and the Soviet Union in 1972.

Commercial communication satellites began in April, 1965, with Intel I Satellite for Comsat (Communications Satellite Consortium), in which 83 nations own stock. The network has 42 ground stations in 29 countries and leases 4000 full-time circuits.

In 1972, the Queen Elizabeth II was joined, by Intelsat IV over the equatorial Atlantic, to the Comsat lab in Maryland, providing the possibility of greater telephone communication and high-speed facsimile service to and from any point in the world and to computers.

The weather satellite, Tiron I, was launched in April, 1960, and transmitted data on the organization of weather systems. The first operations meteorological satellite, established in 1966, was followed by several camera-equipped mechanisms which recorded the world's physical environment and relayed an unbroken flow of meteorological data.

Weather satellites tracked Hurricane Agnes in June, 1972, from its development off the Yucatan, Mexico peninsula. The first satellite data indicated tropical storm intensity June 16, and hurricane intensity June 17, 18, and 19. After coming inland along the Florida panhandle and losing hurricane winds, Agnes was on satellite pictures as the storm moved northeast across Georgia, South Carolina, out along the Eastern Seaboard, and then back inland over Pennsylvania, New York, and the Great Lakes area, causing severe flooding of the Susquehanna River basin in several states. The observed path is given in Figure 29.

The weather satellites have saved lives in the prediction of storms. At the turn of the century, a storm in Texas killed 5000 people, and, in 1959, Mexico was hit by a storm costing 1500 lives. Hurricane Agnes killed 118 people after the storm had lost intensity but was still producing heavy rains inland; these people were lost in floods.

Furthermore, the same period has brought us to an understanding of the power of the atom, in the atomic bomb and the atomic power plant, as well as in its peaceful uses, and particularly medicine's fight against disease.

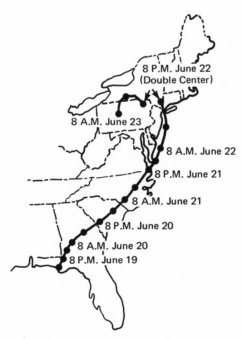

FIGURE 29. Weather satellites plotted the course of Hurricane Agnes in June, 1972.

Medicine, starting with Pasteur's germ theory of disease, has developed enzymes, vaccines, specific remedies, and the appreciation and expansion of public health, sanitation, and nutrition. The practitioners of medicine now understand genetic theory, the chemical composition of nucleic acids, and the biochemical control of life processes.

The twentieth century has shown our highest achievement to date in humanizing the physical forces of the earth. There is now a beginning of the control of climate, with the development of air-conditioned homes, offices, shopping centers, sport stadiums, and transportation vehicles and systems. These are operable at the push of a button by automation engineering.

This humanization has come about by technological development, the specialization into many branches of science, engineering, and industrialization. There are more trained specialists of many diverse types now living and practicing than ever existed during man's recorded history up to 1900. As a result, our knowledge doubles in a decade. With this has arisen the management and coordination of the libraries of the world, and of the memory banks of computers, with their great capacity to employ intricate data, calculations, and coordination to solve problems and their ability to analyze records.

The United States has achieved an immense highway program from coast to coast of many thousands of limited-access roads, freeways, and turnpikes without traffic lights. Engineers have leveled hills and mountains, filled valleys, drilled tunnels, crossed rivers, streams, and arms of the sea for commerce and often just for our pleasure.

The production of food is the task of the largest single group in the most highly industrialized nation in the world, the United States. According to the U.S. Department of Agriculture, farming and food preparation employ nearly 5 million workers on 3 million farms. Each farm worker produces food, fiber, and other commodities for himself and 42 others. An average American consumes 230 lb of meat and poultry, 540 lb of fruit and vegetables, and 580 lb of dairy products annually.

People desire more than simple transformation of the wilderness to farms, more than hospitable, healthy areas, more than pleasure developments, homes, palaces, museums, cities, and manufacturing areas. We want, and must manage, our environment, be it artificial, or the cultivated wilderness of canyons, mountains, reclaimed deserts, or still other unbalanced creations of the earth's natural physical forces.

Humans also want to satisfy their aesthetic sensibilities. We spend our daily lives protecting, developing, and existing within artificial environments that are nonetheless part of larger natural environment, except for those that are too cold, too hot, too dry, too wet, too inaccessible, or too high for habitation. Even the latter regions have been lived in often and long enough for us to study and, to an extent, utilize them.

The twentieth century has brought the new engineering of the astronaut, weightlessness, lunar travel and exploration, and travel by rockets to study and send back instrumental data from Mars and Venus. We now possess space laboratories that circumnavigate the earth, photographing its furthest reaches and teaching us about the planet as a whole.

We are better able to manage the earth through knowledge gleaned from the Earth Resources Technology Satellite (ERTS). Its imaging systems seek and transmit information concerning agricultural and forest resources, mineral and land resources, water resources, marine resources, and land use. This satellite has been in orbit since 1972.

ERTS flies 920 km above the earth in a circular, nearly polar, orbit. It orbits the earth 14 times per day; each pass covers a width of 185 km with overlap. After 18 days, the satellite returns to the first position, having covered the globe. Most of the data from the satellite have come from a multispectral scanner subsystem that views an area 185×185 km in four wavelength bands: the green (500–600 nm), red (600–700 nm), near-infrared (700–800 nm), and a second infrared (800–1100 nm). Water is relatively transparent in the green, but appears black in the infrared wavelength. Vegetation reflects extremely well in the infrared, and is as bright in that wavelength region as snow is in the visible region. The brightness of vegetation in the infrared wavelength depends on the type of vegetation and the health of the plants. Healthy crops appear brighter than diseased vegetation.

The images obtained in the various wavelength regions are transmitted directly to earth when the satellite is over the United States. At other times, the images are stored on magnetic tapes. The individual images can be combined to form artificial color composites. Investigators have become skilled in interpreting these composites.

Repetition of the imaging every 18 days is particularly valuable, since comparison of succeeding images can reveal significant changes. Resolution is limited to about 100 m, and the image as received is distorted. Geologists have been able to identify features that previously had escaped detection.

Many countries have displayed enthusiasm for the ERTS images. Canada is operating its own receivers. Brazil, Mexico, and Venezuela are moving toward establishing their ground stations. The Brazilians are enthusiastic about ERTS, for it is giving them a first look at much of the Amazon Valley.

Poverty, middle class affluence, and the leisurely rich, are purely relative social descriptions, and therefore often totally misleading. Today's average citizen in an industrially advanced country has more mechanical, electrical, electronic, and automated slaves at his service than any pharaoh, conquerer, king, or dictator who ever lived. Their lives were relatively short. Today, as a direct result of our broadly expanded technology, our life-spans are constantly increasing and, indeed, life expectancy in the industrialized nations has doubled during the last 100 years. We are now much more confident than before of being able to manage the physical forces of the earth to satisfy our needs and those of future generations. And this confidence gives us less reason to respect those whom we see as doomsday prophets, those who have so often been completely mistaken in the past. The human race enjoys dominance over, and has great hope to stabilize in perpetuity, the small biosphere on the surface of this turbulent, fragile water planet we call and make our lifetime home.

INDEX

African Game Lands, 180, 182, 183
Age of earth, 2
Agricultural engineering, 147, 148, 149
Alaska, 30,171
Altiplano, 16
Anchovies, 113
Andes, 13
Andromeda nebula, 1
Atacama, 69
Atmosphere, 87, 89
  composition 87
Avalanches 208
Avery Island, La., 197

Baja California, 171
Barometric pressure, 88
Beaches, 103, 107
Beaufort scale, 91
Beryllium, 143
Biosphere, 2
Black Hill floods, 60, 61
Blister rust, 162, 165
Blue earth, 29
Borlaug, N., 151
Building rocks, 38

Cadmium, 143
Calcium, 143
California
  beaches, 107
  state parks, 179
Camille, 91
Carbon cycles, 124, 125, 126
Carbon dioxide, 89, 123, 207
Carbon monoxide, 127, 128, 129, 207
Cellulose, 131

Chemical cycles, 123
Chestnut blight, 163
Chinooks, 100
Chlorine, 144
Climate and agriculture, 79
Coal, 31
Coastal zones, 110, 111
Cobalt, 143
Communication, 214
Composition of earth, 2, 3, 36, 37
Concentration of elements, 37
Conservation, 171
Continental drift, 7, 9
Core, 2
Corn, 152
  corn belt, 64
Cow farming, 200
Creep of glaciers, 68
Crust, 2
Cyclones, 92

Dams, highest, 50
DDT, 162
Deep sea drilling, 12
Deserts, 69, 70, 71
Detritus, 5
Dikes, 117
Discontinuity, 2
Drainage acts, 52, 53, 54, 65
Drainage of wetlands, 61
Dust bowls, 97
Dwarf mistletoe, 165

Earthquakes, 19
  major, 24
  Mexico, 26
  San Francisco, 27

Earth Resources Transmission Satellite
  (ERTS), 217
Elevation of lakes, 48
Elm beetle, 164
El Nino, 60, 110, 113, 115
Elsinore, 192
Energy source, the sun, 2, 4
Engraver beetles, 160
Ergs, 72
Everglades, 194

Feather River, 191
Fiji, 32
Fires, 167, 169
Floods
  control, 51, 52, 56
  Mississippi, 59, 60
Foens, 100
Forests and insects, 157

Ganats, 72
Geologic origin of lakes, 48
Geysers, 80, 84
Glaciers, 67
  creep of, 68
Golden Gate National Recreation Park, 172
Grand Canyon, 171
Great Lakes, U.S., 48
Green revolution, 151
Gypsy moth, 163

Haboobs, 101
Heart rot, 166
Highest mountains, 13, 14, 15
Honey Lake, 189
Horseless Age, 201
Humanizing of earth, 147
Hurricanes, 90, 93, 96
  Agnes, 201
Hydroelectric plants, 51

Insecticides, 153, 155
Interfaces, 103
Irrigation canals, 77
Islands, 103

Kaleheri, 66
Karst regions, 39, 40
Kenya, 181
Kilamanjaro, 31
Kilauea, Hawaii, 29

Krakatoa, 31

Lakes
  elevation of, 48
  geologic origin of, 48
  large, of the world, 49
  largest man-made, 49
Lakes and parks of California, 184, 185,187
Large cities, 207
Las Trampas ridge, 178
Lead, 143
Lethal yellowing, 163
Light, speed of, 1
Light year, 1
Lima, Peru, 16
Lithosphere, 5
Loo, 98
Long Island beaches, 109

Major engineering feats, 210
Major Islands, 104
Malthus, M., 151
Managua, Nicaraugua, 23
Manganese, 143
Mantle, 2, 7
Marshes, 117
Methane, 130
Monsoons, 97, 98, 99
Mountains
  building of, 16
  highest, 13, 14, 15
  highest in U.S., 18
  Mt. Kamai, 31, 35
  Mt. Pelee, 31
Muir, 172

National forests, 171
National parks, 171
National Park Service, 174, 177
Natural gas, 130
Nitrogen, 138, 139
Nullabar Plain, 40

Oceanography, 12
Oil seeps, 133, 134, 135
Okefenokee, 66
Oroville, 190
Overpopulation, 205

Panama Canal

Pangaea, 11
Pesticides, 149, 153
Pests and diseases, 149
Petroleum, 132
Phosphorus, 142
Photosynthesis, 89, 123
Pine bark beetles, 159
Pipe lines, 131, 137
Plagues, 153, 154
Plant foods, 148, 149
Pollution by autos, 201, 202
Pompeii, 32
Portland cement, 39
Prediction of earthquakes, 21
Pressure effects, 4

Rainfall, 59
Reclamation, 74, 78, 79
Red tides, 118
Redwoods, 175, 176
Reefs, 118
Rivers, 51

Sahara, 73
San Andreas fault, 21, 22
Sand bars, 108
Santa Anas, 100
Seismology, 19, 20, 21
   history of, 20
Shasta Lake, 189
Siderosphere, 2
Silicon, 145
Snowstorms, 58
Solar evaporation, 209
Solar system, 1
Speed of light, 1
Spruce beetle, 157
Spruce budworm, 158
Structure, 7
Subsidences, 119, 120
Sulfur and sulfur dioxide, 138, 139, 140,141
Sun
   composition of, 4
   energy source, 2, 4
   plasma, 4

Swamps, 116

Tanzania, 181
Technological growth, 209
Tecolate tunnel, 193
Tectonics, 7, 8, 10, 11, 16
Temperature effects, 4
Tennessee Valley Authority, 55
Tornadoes, 94, 95
Troy, Greece, 203
Tunnels, 211
Tussock moth, 160, 161
Typhoons, 89

Uganda, 181
U.S. geographic facts, 112, 113

Valley of Ten Thousand Smokes, 29, 35
Vesuvius, 12
Volcanoes, 28, 33, 34
   ring of fire, 15, 32
   submarine, 29, 30, 31
Vulcanology, 28

Wallace, Henry, 152
Waste in an urban environment, 199
Water cycle, 41
Water distribution, 42, 43, 80, 82, 83, 84,
   85
Waterfalls, 54, 55
Water transport, 43, 45, 46, 47
Weather, 90
Western pine beetle, 160
Wetlands, 61
   drainage of, 61
White dwarfs, 4
Winds, 97

Yosemite, 172, 173

Zambia, 181
Zinc, 143
Zonda, 100